Flying visit of Truth to
Berlin
in the form of an R.A.F. leaflet raid here fancifully depicted
— but not forgetting a great many hard facts —
by Rex Whistler
STETTINER STATION
HOHENZOLLERN MUSEUM MONBIJOU PALACE
THE CATHEDRAL
THE PALACE
THE TOWN HALL
Lustgarten
LINDEN
FRIEDRICH STRASSE
THE FRENCH CHURCH
CROWN PRINCE'S PALACE
KONIG STR.
GERTRAUDT STR.
GENDARMENMARKT
STRASSE
ALLIANCE
SS
MEIN KAMPF

WAR MAP

PICTORIAL CONFLICT MAPS 1900-1950

WAR MAP

PICTORIAL CONFLICT MAPS 1900-1950

By Philip Curtis & Jakob Søndergård Pedersen

SIFTON, PRAED & Cº. PUBLISHING

— The Map House —

54, Beauchamp Place, LONDON, SW3 1NY

Our grateful thanks to everyone at The Map House for all their help and support: Mary Beal, Vlada Belokon, Clare Bright, Laura Dadswell, Jessie Fahy, Selene Obolensky, Charlie Savile, Jacki Scrips and Peter Stuchlik.

A special thank you to Peter Barber for his invaluable insights and suggestions.

For Tony Savile, who had the foresight to recognise
the historical and artistic value of 20th-century maps,
and whose encouragement over so many years led directly to this book.

THE MAP HOUSE

Established 1907

The Map House 2016

First published in Great Britain in 2016 by
The Map House
Sifton Praed & Co Ltd
54 Beauchamp Place
London SW3 1NY

www.themaphouse.com
maps@themaphouse.com
+44 (0)207 589 4325

ISBN: 978-0-9571497-1-7

Written by Philip Curtis & Jakob Søndergård Pedersen.

Design and production by Jamm Design Ltd.

Printed in Great Britain.

Perse
ASIE
Brésil
A Robida

FREEDOM

Punch Magazine, 6 November 1939

Foreword

War Map is perhaps something of a misnomer for a book which vividly illuminates world politics as well as military affairs, and which could be read with profit by anyone interested in twentieth-century culture as well as politics. Side-by-side with its images, it contains information that will be new to most of its readers – for instance about the *Propaganda-Abteilung Frankreich*, through which the German occupiers sought to mobilize French opinion against the Allies during the Second World War. The maps enable the reader to view the problems of the past through the eyes of some of the people who lived through and had to make sense of them for others. Strikingly modern though many of the images are, they had their roots in the past.

War and political maps intended to influence onlookers, go back to classical antiquity at the very least. It has been argued by Richard Talbert that the so-called Peutinger Table, generally accepted to be a copy dating from about 1200 of a fourth- or fifth-century Roman map of its Empire, was intended as propaganda; it was originally displayed as a frieze above the throne in the audience chamber in the palace of one of the emperors.[1] It is widely believed too that around the time of Christ's birth, a world map, also showing the extent of the Empire and commissioned by the Emperor Augustus's son-in-law Marcus Vipsanius Agrippa, was commissioned for display on a wall along the processional route that led to the Roman Forum.[2] Political propaganda maps reappeared during and after the European Renaissance, sometimes in the form of apparently objective maps, where the propaganda message was conveyed by colouring and border decoration, as seen in an early eighteenth-century map of the Austrian Habsburg dominions and later in maps of the British Empire.[3]

Gillray (James): The Plumb-Pudding in Danger, 1805

Crane (Walter): World, 1886

In late nineteenth- and early twentieth-century political maps, colour was used not only to distinguish between the landmasses of sovereign states and empires but also, implicitly, to make moral points. Allied mapmakers, for instance, used black to denote Nazi Germany (*Nazi War Aims:… Grab, Grab, Grab,* 1939; *Zog it off,* 1939; *The World Divided,* 1942) just as British map publishers usually used red to highlight and to generate a somewhat aggressive pride in the British Empire, particularly when it appeared to be threatened. An example is the map *The Overseas Club: How the World is at War* in 1916, where red is also used to with similarly aggressive connotations to highlight German expansionism in the Maurice Neumon map, *La Guerre est l'Industrie Nationale de la Prusse* of 1917.

[1] Richard J.A. Talbert, *Urbs Roma to Orbis Romanus: Roman Mapping on the Grand Scale* in Richard J.A. Talbert, ed., Ancient Perspectives. *Maps and Their Place in Mesopotamia, Egypt, Greece & Rome* (Chicago and London, University of Chicago Press, 2012), pp. 177-90.

[2] Talbert, pp. 167-72.

[3] Johann Christoph Homann, *Tabula Geographica Europae Austriacae Generalis…*, (Nuremberg: J.C. Homann, ca. 1730); [John Colomb & Walter Crane], Map of the World showing the extent of the British Empire in 1886 (London: the Sphere, 1886).

Covens (Johannes and Cornelis Mortier): Grand Theatre de la Guerre..., circa 1730

Propaganda war and campaign maps had a somewhat different history. It is possible that the so-called Dura-Europos map – the fragment of a map of the Black Sea on the remains of a mid-third-century Roman shield now in the Bibliothèque Nationale de France – may originally have commemorated a military campaign or campaigns.[4] Seventeenth-century rulers such as Louis XIII, Louis XIV, and their allies and enemies were prepared to expend considerable sums on commissioning magnificent figurative wall maps of particular battles and sieges from outstanding artists, such as Callot or Romeyne de Hooghe.[5] Campaign maps, often called *Theatre of War*, became relatively common in Europe from 1600 but had their predecessors in the printed maps tracing the course of the wars between the Germans and the Turks in the previous century. In their most prestigious form these were compilation maps, assembled by geographers from what they judged to be the best sources, and adorned around the borders, and occasionally within the map itself, with appropriate figurative decoration.[6] More often, wars presented unscrupulous map publishers with the chance to profit at minimum cost to themselves from the sudden widespread public thirst for knowledge about the course of a campaign in well-mapped regions, in which their nearest and dearest were involved. Out-of-date and otherwise unsellable copperplate maps of the relevant areas were recycled and disguised by re-titling or, occasionally, cut up and re-assembled, to create an apparently new whole, sometimes adorned with freshly-engraved and appropriate imagery.[7] Early twentieth-century political and military maps participated in this heritage. As Philip Curtis and Jakob Søndergård Pedersen point out, for instance, Ernest Dudley Chase reused large elements of his *Total War Battle Map* of 1942 to serve the purposes of the Cold War in his *Factual and Pictorial Map of World Freedom* eight years later.

[4] Benet Salway, *Putting the World in Order* in Talbert (see note 1), pp. 196-7; plate 9.; Richard Talbert, *Celebrating the Roman Black Sea*, in Peter Barber (ed.), The Map Book (London: Weidenfeld & Nicolson, 2005), pp. 24-5.

[5] E.g. Jacques Callot's map of the Siege of Breda (1627-8) illustrated and discussed in Barber & Harper, pp. 68-9; Romeyne de Hooghe, Luxemburgum (Amsterdam: N.Visscher, 1684) illustrated and discussed in Peter Barber, *A French Triumph* in The Map Book (see note 4 above), pp. 178-9.

Military and political maps, particularly those created before 1920, were often enlivened by figures of people and animals. The beginnings of the figurative tradition can be traced back to the Middle Ages. In the early fourteenth-century an Italian priest, Opicinus de Canistris, transformed the geographical outlines on Mediterranean Sea, or portolan charts, into people. He used them to highlight theological and moral points about contemporary society.[8] Between the sixteenth- and eighteenth-century, in a similar way, maps were moulded to create the depiction of Spain and its empire as a queen, or the Netherlands as a lion.[9] The intention seems to have been to evoke patriotic pride and a sense of solidarity. In the nineteenth-century the use of figures was occasionally linked to geographical outlines, as in the work of Lilian Lancaster, the creator of the original *Geographical Fun* maps published in 1868, which were long and wrongly attributed to Aleph, the author of the text.[10]

Van den Keere (Pieter): Leo Belgicus, 1617

[6] See John Elphinstone's manuscript *New Map of North Britain* (1746) commemorating the Jacobite uprising of 1746-7 Peter Barber & Tom Harper, Magnificent Maps. Power, Propaganda and Art (London: British Library, 2012), pp. 110-111.

[7] Dirk de Vries, *Dutch Cartography* in Robert P. Maccubbin & Martha Hamilton-Phillips (eds.) The Age of William III & Mary II. Power, Politics and Patronage 1688-1702. A reference encyclopedia and exhibition catalogue (Williamsburg: The College of William and Mary; New York: Grolier Club; Washington D.C.: The Folger Shakespeare Library; 1989), pp. 109-110.

[8] Karl Whittington, Body-Worlds Opicinus de Canistris and the Medieval Cartographic Imagination (Turnhout: Brepols, 2014).

[9] Heijden, H.A.M. van der Leo Belgicus An illustrated and annotated carto-bibliography (Alphen aan de Rijn: Canaletto, 1990); Gillian Hill, Cartographical Curiosities (London: British Library, 1978), pp. 38-63.

[10] See http://barronmaps.com/lilian-lancaster-1852-1939/

Lancaster (Lillian): Italy, 1886

Increasingly stock national stereotypes, such as John Bull and animals, were exported onto maps. This fashion ultimately derived from the peoples and sometimes fictionalised creatures shown on medieval European *mappaemundi* such as the Hereford and Ebstorf world maps of around 1300, where they represented moral attributes as well as providing what was thought to be factual information about the peoples and history of the world.[11] The tradition continued through the sixteenth-century, as epitomized by the decorative charts of Catalan and, later, Italian portolan chartmakers and by the splendid maps of the Dieppe School that included John (Jean) Rotz and Pierre Desceliers.[12]

[11] See Sarah Arrowsmith, Mappa Mundi. Hereford's Curious Map (Woonton Almeley: Logaston Press, 2015), pp. 43-78.

[12] Helen Wallis, The Maps and Text of the Boke of Idrography presented by Jean Rotz to Henry VIII now in the British Library (Oxford: OUP for the Roxburghe Club, 1981). (ed.); Chet van Duzer, The World for a King. Pierre Desceliers' Map of 1550 (London: British Library, 2015).

Vicus (Johannes Maria): [West Africa], 1597

Rose (Fred): Serio-Comic War Map, 1877

Though people and animals are to be found from about 1848 on mass-produced satirical political maps, from 1877 Frederick Rose magnificently exploited their popular attributes, or, in the case of the octopus, their physical form, as an allegory for the policies and ambitions of the nations that they represented. Such was Rose's influence that, as is demonstrated in this book, the octopus, first associated with Russia in Rose's cartoon of 1877, was used up to 1945 to represent, and cast aspersions on, Imperial Germany, Winston Churchill's Britain, the capitalist USA and a militarist Japan. It still continues to put in brief appearances in more recent newspaper cartoons.

Girolamo Savonarola, 1497

The emotive power of maps as symbols led them to feature as propaganda on medals as early as the late fifteenth-century[13] and in political cartoons from at least the early eighteenth-century. Perhaps the best-known early cartographic cartoon is Gillray's brilliant *The Plumb Pudding in Danger* of 1805, showing Napoleon and William Pitt partitioning the globe, depicted as a Christmas pudding. Gillray may, however, have got his idea from a cartoon of 1772 satirising the first partition of Poland. This was marketed in England as The Royal Cake and showed the rulers of Russia, Austria and Prussia tearing apart an up-to-date map of Poland, to the despair of its king.[14]

Dutch wall maps with decorative borders, the so-called *cartes-á-figures* that first appeared in the 1590s and their successors, the splendid wall maps produced in Europe in the seventeenth- and eighteenth-centuries, were another clear influence on the format, though less on the content, of several of the twentieth-century political and war maps. Though highly decorative, their publishers selected the border illustrations – depictions of the elements, town views, fashion plates, portraits, wild life and the occasional battle scene – to make cultural, political and patriotism points.[15] The influence can be clearly seen in the American *Historical Map* of 1932, the *Pictorial History and World Map* of 1939 and the *Defense Map* of 1941, where the imagery may be updated to include flags, battle ships, bombers and weaponry but the patriotic purpose remains the same.

De Fer (Nicolas): Asia, 1698

[13] G.F. Hill & Graham Pollard (eds.), Renaissance Medals from the Samuel H. Kresss Collection at the National Gallery of Art (London: Phaidon, 1967), no. 282 a bronze medal of about 1497 illustrating a prophecy of the Florentine friar Savonarola and showing the dagger of the Lord, representing the French invasion, above a Ptolemaic map of Italy.

[14] Peter Barber. *Paper Partition*, in The Map Book (see note 4), pp. 226-7.

[15] Günter Schilder, *Jodocus Hondius, Creator of the Decorative Map Border*, The Map Collector 32 (September 1985), pp. 40-43; Günter Schilder, Monumenta Cartographica Neerlandica (Aalphen-an-den-Rijn:Canaletto, and (vol 9) Hes & De Graf. 9 vols. 1986-2013).

Sylvanus (Bernard): [World], 1511

Twentieth-century war maps have also played with projections and orientations for depicting regions of the world with a freedom which has not been seen since the sixteenth-century. Before the Mercator Projection and, allied with it, the North orientation began exercising an iron grip over the way in which most people viewed the world, mapmakers experimented with projections and orientations. One of the most intriguing is the so-called cordiform or heart-shaped project, particularly associated with Philip Apian, Bernard Sylvanus, Oronce Fine, the young Gerard Mercator and Abraham Ortelius working in the mid-sixteenth-century. This projection, in which the world resembles a heart, is not unlike the ones favoured by Richard Edes Harrison and his followers as a way of charting the course of the Second World War. But where for Harrison it was primarily a means of aiding understanding of and dramatising military and political strategies, Giorgio Mangani has argued that for Apian, Fine, Mercator and Ortelius it was a reference to their religious beliefs, and in particular a group know as the Family of Love which had the heart as its symbol. [16]

Occasionally the twentieth-century maps reveal the specific cartographic sources which their creators consulted beforehand. The format, scale cartouche and compass star of Tom Luftin Johnson's 1944 map of West Point, for instance, strongly suggests that he had looked at one of the manuscript maps by late sixteenth-century Portuguese hydrographers and particularly Fernao Vaz Dourado, while the depiction of European rulers on Kampf's map of Europe in 1915 owe something to their portrayal in the finer manuscript atlases of the Venetian sixteenth-century chartmaker Battista Agnese.[17]

For all the inevitable influences from the past, however, the most striking aspect of the maps produced from about 1930 is their novelty, which owes nothing to historic precedent. The bold colours and clear lines, and their unusual orientations, particularly associated with Richard Edes Harrison, all display the influence of Art Deco and the popular poster designs of the 1920s and 1930s. They also reflect major underlying changes in technology and society. It would have been impossible without the increasing familiarity with flying, and with aerial photos of large areas, to create images which far surpassed traditional bird's-eye views and the supposed 'balloon views' of the nineteenth-century. Familiarity with views from planes must also have encouraged mapmakers to dispense, at times, with what by 1900 had become the fetish of orientation to the north. The new cartography also owed much to improved colour-printing techniques which allowed stronger colours and sharper outlines. Above all, mass production, mass marketing and mass markets led to mass audiences for maps – audiences familiar with posters and advertising but not with traditional maps. This audience – the international mass public – had few preconceived, traditional ideas about the nature of mapping and a consequent readiness to accept and even embrace the new – which often seemed exciting as well as enlightening.

This book presents some of the most striking images to be created in the first half of the twentieth-century. The high quality of the reproductions enhances the reader experience. But by combining them with a very informative and well-referenced text accompanied by telling and well-reproduced details, Philip Curtis and Jakob Søndergård Pedersen have created a work which will continue to be consulted for years to come by anyone interested in the art of twentieth-century graphic communication.

Peter Barber, OBE, FSA, FRHistS
President of the International Map Collectors' Society
Head of Map Collections at the British Library, 2001-2015.

[16] Giorgio Mangani, *Abraham Ortelius and the Hermetic Meaning of the Cordiform Projection*, Imago Mundi 50 (1998), pp. 59-83.

[17] For an illustration of the work of Vaz Dourado see Monique de la Roncière, Michel Mollat du Jourdin, Les Portulans. Cartes marines du xiiie au xvii siècle (Fribourg: Office du Livre, 1984), p. 59; Arthur Dürst (ed.) The Portolan Atlas of 1546 of Battista Agnese from the Russian National Library St Petersburg (Graz: Akademische Druck- und Verlagsanstalt, 1993).

W.F.B.

Introduction

The first half of the 20th-century was an unprecedented period of change in the history of mankind, and nowhere was this transformation more apparent than in the field of conflict.

The most important and far reaching of these changes began in the nineteenth-century and would change the very nature of war itself. This was the steady replacement of animal-powered transport by the railway. The fundamental effect was to allow armies, no longer reliant on the local supply chain, to increase vastly in size.

The immense conscript armies now needed to fight a 'modern' war required a level of tacit support from the societies from which they were drawn, that would in turn transform the printing and use of propaganda.

The need for nationwide backing for the war effort, at a time when ordinary people had unprecedented access to a multitude of different and often contradictory information sources, made effective domestic propaganda essential.

It was the new form of 'industrialised' war, with its insatiable need for a continuous stream of conscripts to feed into the fray, that led to the massive boom in the printing and distribution of all forms of propaganda material.

After defeat in the First World War, it was widely held in Germany that it was the collapse of morale on the home front, rather than military failure, that was the principal cause of their downfall and that this was partly due to inadequate use of propaganda.

While propaganda as a political tool has always been in existence, no one made quite as much use of it as the fascist states during the twentieth century, in both peace and war. In *Mein Kampf*, Adolf Hitler devotes Chapter VI, *War Propaganda,* and Chapter XI, *Propaganda and Organization,* to the description and use of propaganda.

The western democracies responded vigorously to this battle for hearts and minds. The two World Wars therefore provided a vast upsurge in material, with both sides regarding it essential to maintain the morale of the fighting troops and those back home.

Separately issued printed matter was a very popular form of propaganda and was regarded as having a greater reach than film and newsreel. When government was involved and the artist was acting as a national spokesperson, the cartoons, caricatures and maps could be printed in significant numbers. But when published independently, and occasionally even illegally, the wartime restrictions on colour printing and the use of wood pulp would reduce output considerably. This has left a wide range of pieces, but the poor quality of wartime paper means that even those originally printed in quantity often only survive in a handful of examples. Many others are known by a single copy.

A further major limitation of the number of surviving examples is the deliberate destruction of propaganda works created by a defeated faction or nation. It would be a reckless Bulgarian indeed who, with the Soviet Army poised on the border, would risk being found in possession of the viciously anti-communist poster, *What Will Happen to Europe if the Communists Win* (see pp.202-3). Conquering armies also habitually destroy the images and symbols of the vanquished.

The power of pictorial maps to deliver a message clearly and powerfully has long been recognised, but it was not until the twentieth century that this form of image-making came fully into its own. The innovation of 'global warfare' fought worldwide made maps an essential instrument in illuminating the ever-changing situation.

As the historian Nicholas Spykman states: *"Geography is the most fundamental factor in foreign policy of states because it is the most permanent. Ministers come and go, even dictators die, but mountain ranges stand unperturbed. George Washington, defending thirteen states with a ragged army, has been succeeded by Franklin D. Roosevelt with the resources of a continent at his command, but the Atlantic continues to separate Europe from the United States and the ports of the St. Lawrence River are still blocked by winter ice. Alexander I, Czar of all the Russias, bequeathed to Joseph Stalin, simple member of the Communist Party, not only his power but his endless struggle for access to the sea, and Maginot and Clemenceau have inherited from Caesar and Louis XIV anxiety over the open German frontier."*[18]

The majority of the maps described here are ephemeral pieces. They were produced to reflect a particular idea or in reaction to a specific event. Their very ephemeral nature is an essential part of their interest and ultimately importance. Unlike some 'high' art which may be regarded as timeless, each of these maps is wholly a product of its period. It is this immediacy which allows them to shine a spotlight on a very precise moment or concept. These evocative flashes of illumination offer extraordinary insights into their moment of history. The power of these images still resonates with us today and allows us to capture these fleeting moments of the past.

[18] Spykman, Nicholas J. 1942, *America's Strategy in World Politics: The United States and the Balance of Power*, New York, pp. xv, 41.

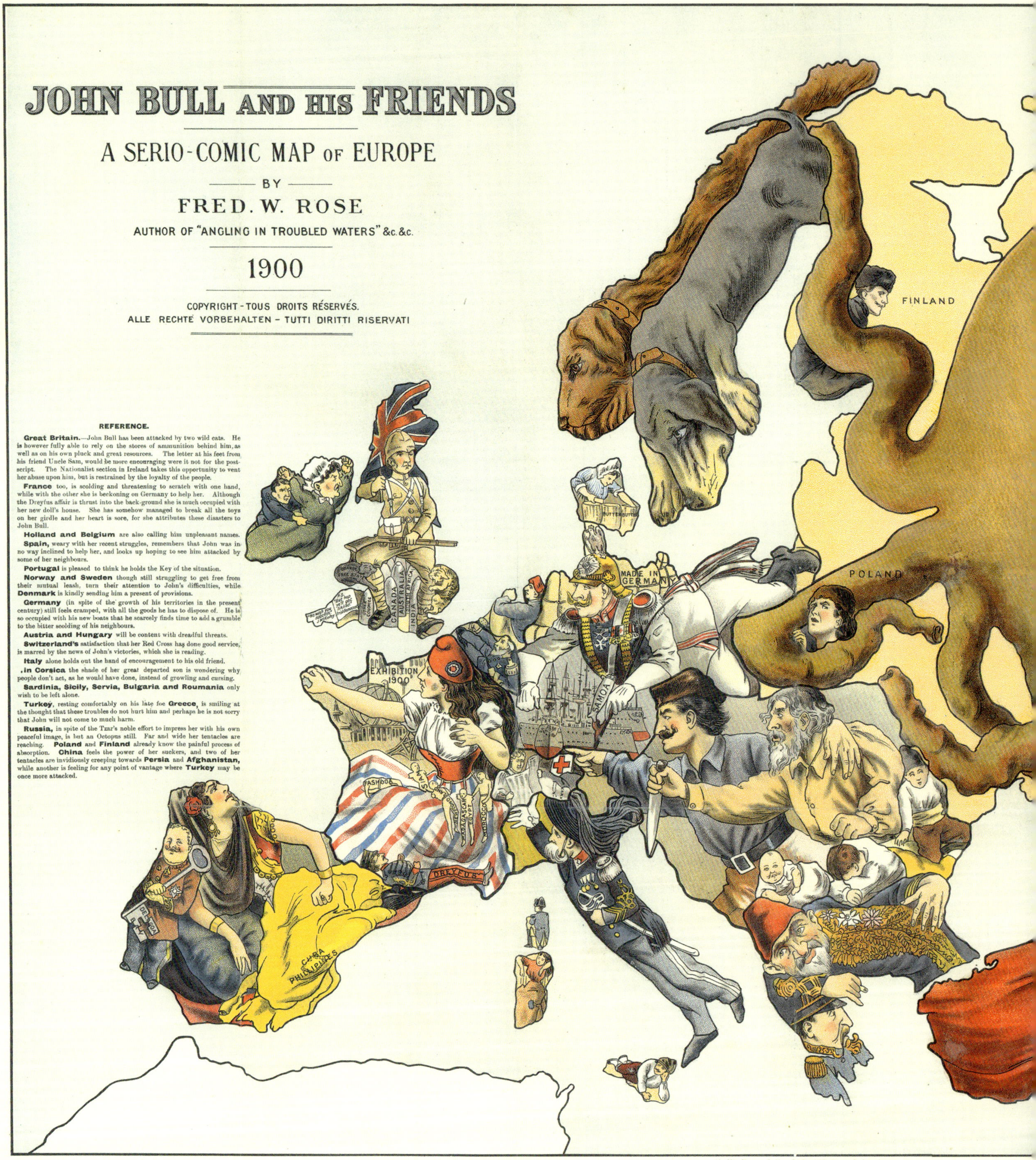

Rose (Frederick W.): John Bull and his Friends

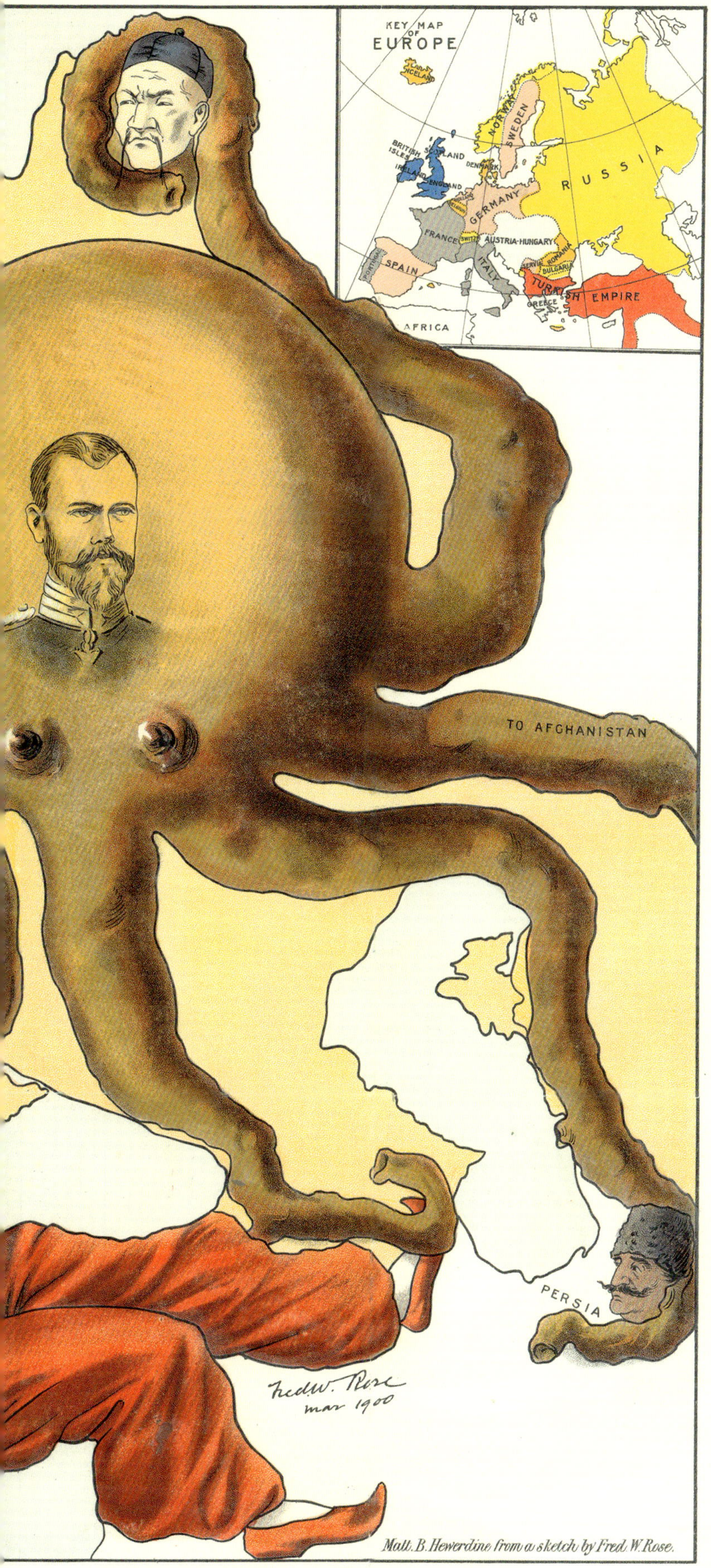

Rose (Frederick W.): John Bull and his Friends

chromolithographed map, G. W. Bacon and Co., London, 1900, 27 x 19 in. (68 x 49cm)

This is one of the great iconic pictorial images of the twentieth-century. The artist, Fred Rose, has managed to capture all the internecine resentments and rivalries of Europe, a continent on the very eve of self-destruction.

Frederick Rose was undoubtedly the most influential of all anthropomorphic mapmakers of the modern age. His first *Octopus Map* of 1877 [see map on p.xiv] set the style of satirical and propaganda mapmaking, and was widely copied throughout the world over the next century. This map is a reworking of Rose's 1899 image *Angling in Troubled Waters* and was drawn in collaboration with the commercial artist, M.B. Hewerdine.[19]

In Rose's 1877 map Russia is portrayed as a wholly malevolent grasping octopus, while here the ferocity of the creature is tempered with a rather more benign portrait of Tsar Nicholas II, the spitting image of his first cousin, the Duke of York, later King George V. This transformation continues in *Hark, Hark the Dogs do Bark!* [see map on pp.18-19] of 1914 where Russia is now depicted as a saviour of the Allies, with the Tsar driving a steamroller and crushing Germany from the east.

[19] Matthew Hewerdine has the distinction of inspiring one of the world's very first film advertisements. In 1894 he was commissioned by Dewar's Whisky to paint *The Whisky of his Forefathers*. In 1898 Thomas Edison used the painting as the basis for a film advertising the whisky. Nead, Lynda, 2007, *The Haunted Gallery*, New Haven, CT., pp. 80-81.

Britain dressed in khaki is preparing to march off to fight in the Boer War. There is a note of encouragement at his feet from J.McKinley, saying "We wish you success" with a discouraging postscript "But hop ye'll get likked." from an Irish Democrat.[20]

[20] Presumably this refers to William (rather than J.) McKinley, the American President assassinated the following year by the anarchist, Leon Czolgosz. The McKinley administration, while maintaining its neutrality during the Boer War, remained broadly supportive of Britain. On the other hand several hundred Irish and Irish-Americans went further than just hoping Britain would be beaten, and actually volunteered and fought alongside the Boers. The Irish Transvaal Brigade was led by Col. John Blake, an Irish-American graduate of West Point.

Britain, France and Germany are all shown variously supported or encumbered by their overseas colonial possessions.

France, represented by *Marianne*, is kneeling on top of the prostrate figure of Captain Alfred Dreyfus, the victim of a gross miscarriage of justice which had split the political left and right in France and was still convulsing the country.[21] It is interesting that in 1900, prior to the signing of the *Entente Cordial* between Britain and France in 1904, the traditional rival France is still perceived as great a threat to British interests as Germany. The mention of 'Fashoda' illustrates this. Now in South Sudan, Fashoda was the site of a major confrontation between the British and the French in 1898 which led to the effective establishment of a French colonial sphere in Central and Western Africa, and a British sphere, including Egypt and the Sudan, to the East.

Germany, represented by Wilhelm II adorned with her most recent colony Samoa, is occupied with building up her fleet of battleships, while her growing industrial might is symbolized by the packages behind the Kaiser labelled 'Made in Germany'.

[21] In 1886 Dreyfus, a Jewish officer, was falsely accused of selling military secrets to the Germans and in a notorious trial dominated by anti-Semitism, was found guilty of treason and sentenced to life imprisonment on Devils' Island. After a dogged campaign of support led by Émile Zola among others, he was re-tried, found guilty once more and sentenced to 10 years imprisonment, and then promptly released. It was not until 1906 that he was at last exonerated. In a final irony Dreyfus himself was the only soldier and major protagonist involved in 'the Dreyfus Affair' to actually see action against the Germans in the First World War. He retired from the army as a Lieutenant Colonel and was made an officer of the Legion d'honneur in 1919. See: Cahm, Eric, 2014, *The Dreyfus Affair in French Society and Politics*, London.

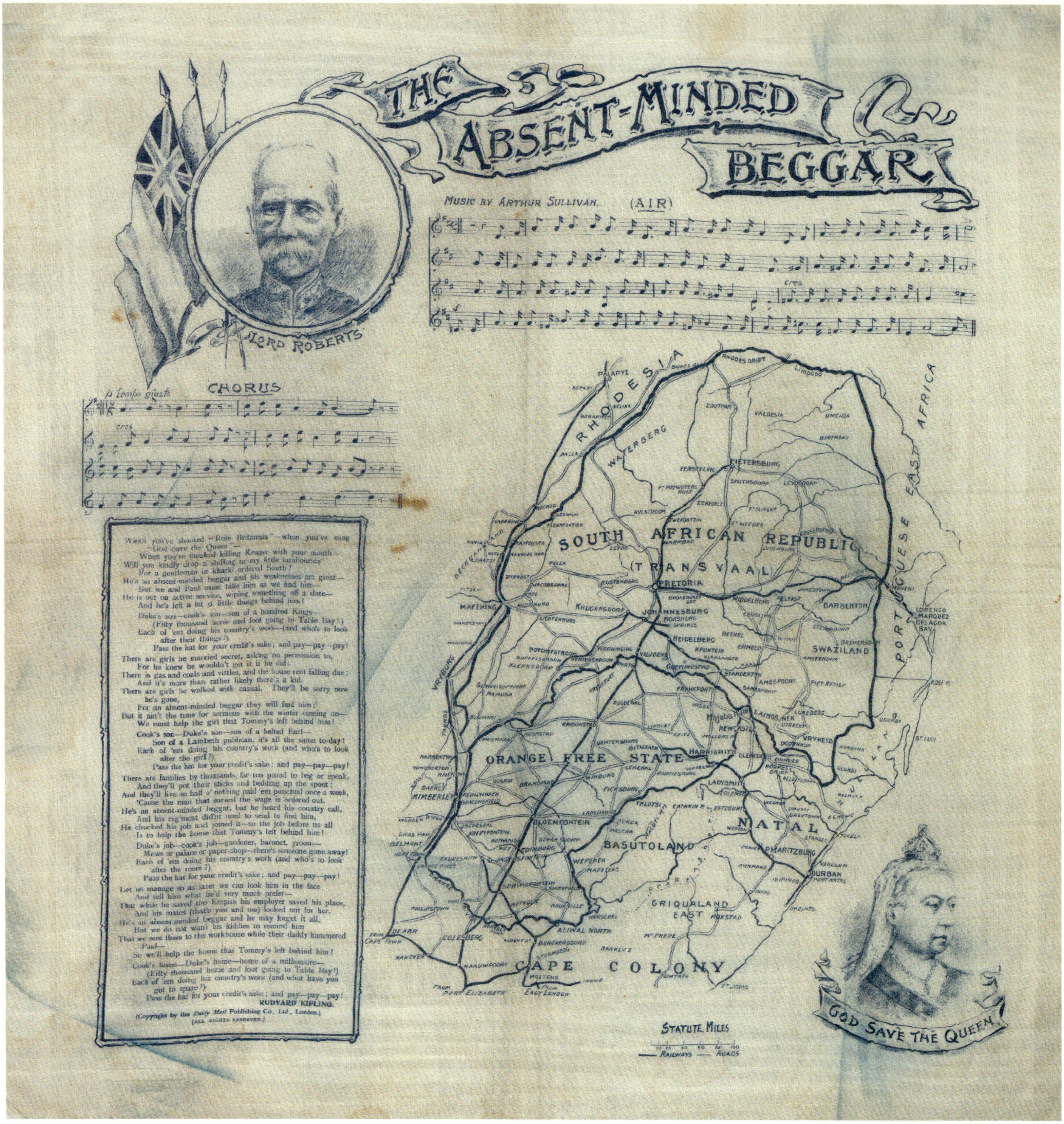

The Daily Mail Publishing Co. Ltd.: The Absent-Minded Beggar

The Daily Mail Publishing Co. Ltd.: The Absent-Minded Beggar

lithographed map printed on cloth, London, c.1900, 18 x 18 in. (46 x 46cm)

This celebrated illustrated verse was designed to raise funds and was written as part of a campaign by *The Daily Mail* to support the morale of the British fighting man during the Boer War.

The song *The Absent-Minded Beggar,* with words by Rudyard Kipling and music by Arthur Sullivan, is one of the greatest charitable money-raising phenomena of all time. The majority of the servicemen called up to fight were ex-soldiers from the Army Reserve. Most were pulled out of civilian employment, with no guarantee of jobs to return to, and no means of supporting their families and dependants other than on an infantryman's salary of 'a shilling a day'.

The release of the song coincided with a wave of fervent jingoism sweeping the country fanned by papers like *The Daily Mail,* and the public response was astonishing. *The Daily Chronicle* wrote: *"It has not been often that the greatest of English writers and the greatest of English musicians have joined inspiring words and stirring melody in a song which expresses the heart feelings of the entire nation."* Its creators were more modest, with Kipling describing it as *"catchpenny verse and I want it to catch just as many pennies as it can…"*. Sullivan wrote to Kipling, *"Your splendid words went with a swing and enthusiasm which even my music cannot stifle."* [22]

Hundreds of thousands of copies of the sheet music were sold; at one point 40 clerks were answering 12,000 requests for copies a day. Both Kipling and Sullivan had waived payment stipulating that all copyright fees should be paid into the charitable fund. The sale of sheet music, performance fees and the sale of merchandise such as this commemorative handkerchief, eventually raised the astonishing sum of £250,000. Today this would be the equivalent of £85,000,000. [23]

Local *Absent Minded Beggar Relief Corps* branches were opened throughout the world, from the West Indies to China. The proceeds of the fund were used to send comforts to the troops, financial relief for their families and the building of numerous hospitals. In December, after the first £50,000 was raised, *The Daily Mail* asserted, *"The history of the world can produce no parallel to the extraordinary record of this poem."* [24] It has even been described, with justifiable hyperbole, as the origin of the welfare state.[25]

[22] Cannon, John, 1997, *The Absent-Minded Beggar, Gilbert and Sullivan News*, March 1997, vol. 11, no. 8, London, pp. 16–17.

[23] In comparison, Band Aid's *Do They Know It's Christmas?,* released in 1984 and one of the most successful singles of all time, raised in the region of £10,000,000 for charity.

[24] *The Daily Mail*, December 1899.

[25] Cannon, John, 1997, pp. 16–17.

'Black Octpus' is a name newly given to Russia by a certain prominent Englishman. For the black octpus is so avaricious, that he stretches out his eight arms in all directions, and seizes up every thing that comes within his reach. But as it sometimes happens he gets wounded seriously even by a small fish, owing to his too much covetousness. Indeed, a Japanese proverb says: "great avarice is like unselfishness." We Japanese need not to say much on the cause of the present war. Suffice it to say, that the further existence of the Black Octpus will depend entirely upon how he comes out of this war. The Japanese fleet has already practically annihilated Russia's naval power in the Orient. The Japanese army is about to win a signal victory over Russia in Corea & Manchuria. And when.............. St. Petersburg? Wait & see! The ugly Black Octpus! Hurrah! Hurrah! for Japan.
March, 1904.
Kisaburo Ohara.
R U
ATLANTIC OCEAN
GREAT BRITAIN
NORWAY
SWEDEN
FINLAND
DENMARK
HOLLAND
BELGIUM
GERMANY
POLAND
FRANCE
AUSTRIA
HUNGARY
SPAIN
PORTUGAL
ITALY
POPE
CORSICA
SARDINIA
SICILY
GREECE
BLACK SEA
CRIMEA
BOSPORAS
TURKEY
CASPIAN SEA
PERSIA
ARABIA
CRETE
MEDITERRANEAN SEA
著作權所有
明治三十七年二月廿三日印刷
仝年仝月廿八日發行
第四版
發賣元 東京市京橋區築地二丁目五番地 臺紙開舍
賣捌所 東京市芝區三田四國町二番地壹號 小西天津館
仝 仝 淺草區南松山町四番地 名書堂 宇田川安高
仝 仝 神田區一ッ橋通町十七番地 田中良三
仝 仝 京橋區南傳馬町三丁目廿一番地 集畫堂森利助
大日本帝國萬歲!!!
明治三十七年三月
東京市京橋區築地貳丁目五番地
著作兼發行印刷者 藪崎芳次郎

Ohara (Kisaburo): A Humorous Diplomatic Atlas of Europe and Asia, 1904

Ohara (Kisaburo): A Humorous Diplomatic Atlas of Europe and Asia

lithographed map, March, 1904, 22½ x 16 in. (57 x 41 cm)

One of the more unusual of the caricature maps derived from Fred Rose's Octopus, this highlights the Japanese fear of Russia's expansionist ambitions in Asia.

It was published at the very start of the Russo-Japanese War of 1904-05. With the text in both Japanese and English, this powerful propaganda image was intended for both domestic and international distribution. The Russian octopus is shown with its tentacles reaching out covetously for Finland, Tibet, Persia, Turkey, Poland and Manchuria.

Her earlier defeat during the Crimean War is indicated by the open wound in the tentacle grasping Crimea. Areas of previous conflicts involving Russia such as Finland, Poland and the Balkans are represented with skulls. Naturally it is the tentacle stretching through Siberia and into Manchuria that most concerned the Japanese. This represents the Trans-Siberian Railway, and its newest section, finished in 1898 running from Harbin to the strategically important harbour of Port Arthur (Lüshun).

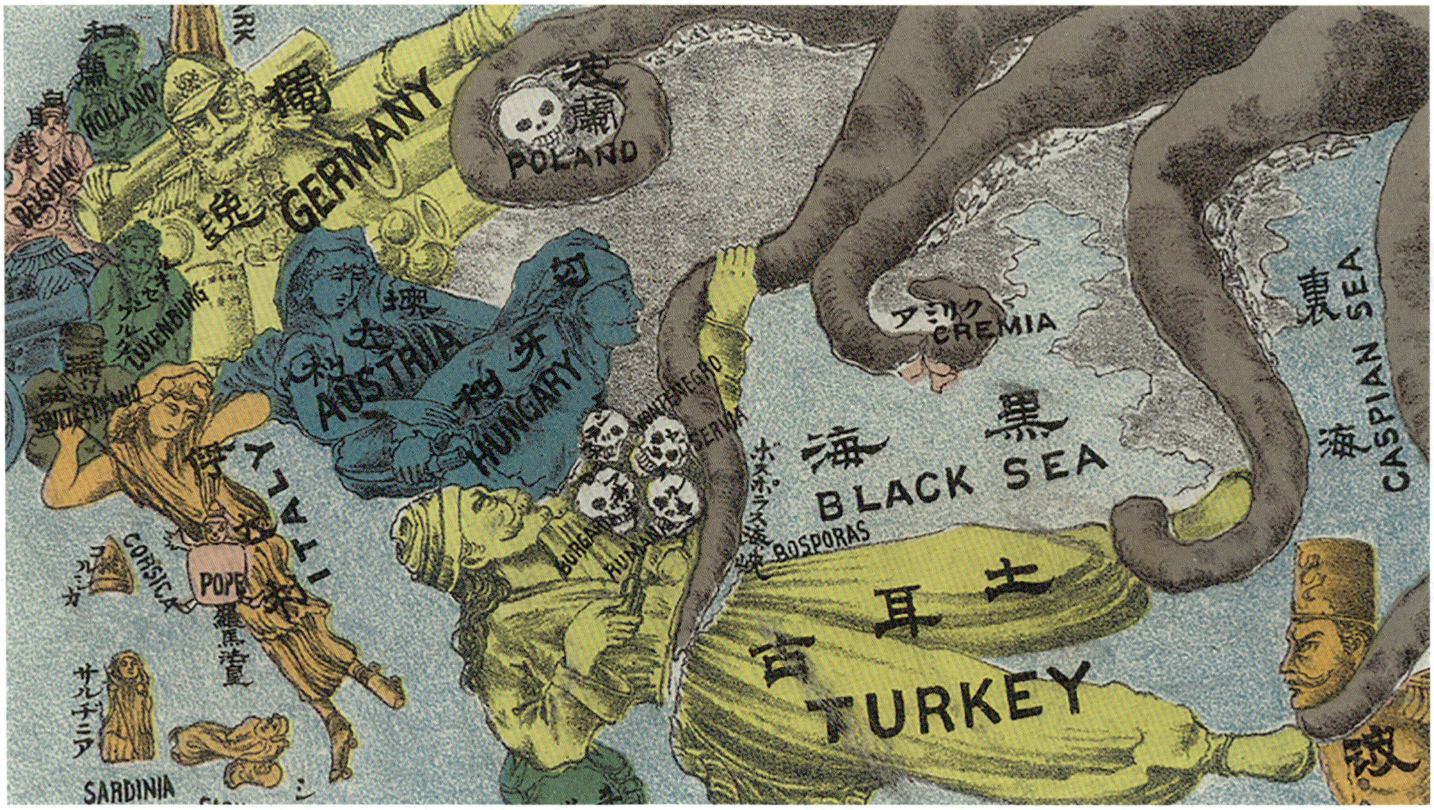

Russian and Japanese expansionist ambitions for both Manchuria and Korea had long put the two countries on a collision course. The flashpoint for conflict was Port Arthur. Russia was desperate for a warm water port on the Pacific, as Vladivostok remained frozen and unusable for much of the winter. Port Arthur was equally essential to the Japanese for their planned conquest of Manchuria. The Japanese first seized the port from the defeated Chinese at conclusion of the First Sino-Japanese War of 1895-96, but were forced to relinquish it under pressure from 'The Triple Intervention' consisting of France, Germany and a self-serving Russia.[26] After extensive newspaper reporting of the Japanese massacre of Chinese civilians during the capture of Port Arthur, Japan received no support from its nominal allies, Britain and the USA. Japan had to watch in impotent fury as the Russians swiftly occupied and fortified the seaport.[27]

The Japanese steadily built up their land and naval forces until they were strong enough to strike back. On 8 February 1904, three hours before formally declaring war, they launched a surprise attack on Russia's Pacific Fleet anchored in Port Arthur.[28]

[26] Gardner, Hall, 2015, *The Failure to Prevent World War I: The Unexpected Armageddon*, Farnham, p. 109.

[27] Creelman, James, 1901, *On the Great Highway, the Wanderings and Adventures of a Special Correspondent*, Boston, chapter V (re-issued online by Cardinal Books: http://www.cardinalbook.com/creelman/highway/iso8859/)

[28] Although similar to Japan's undeclared attack on Pearl Harbor on 7 December 1941, the attack on Port Arthur was not actually against the Laws of War as commencing hostilities before a formal Declaration was only made illegal in 1907 as a direct result of this attack. See: Yale University, *Laws of War: Opening of Hostilities (Hague III)*; 18 October, 1907, Avalon Project at Yale Law School. Available online at http://avalon.law.yale.edu/20th_century/hague03.asp

'Black Octpus' is a name newly given to Russia by a certain prominent Englishman. For the black octpus is so avaricious, that he stretches out his eight arms in all directions, and seizes up every thing that comes within his reach. But as it sometimes happens he gets wounded seriously even by a small fish, owing to his too much covetousness. Indeed, a Japanese proverb says: "great avarice is like unselfishness." We Japanese need not to say much on the cause of the present war. Suffice it to say, that the further existence of the Black Octpus will depend entirely upon how he comes out of this war. The Japanese fleet has already practically annihilated Russia's naval power in the Orient. The Japanese army is about to win a signal victory over Russia in Corea & Manchuria. And when............... St. Petersburg? Wait & see! The ugly Black Octpus! Hurrah! Hurrah! for Japan.

March, 1904. Kisaburo Ohara.

This map, published only a month after the start of the war, declares prophetically that *"The Japanese fleet has already practically annihilated Russia's naval power in the Orient."* This was somewhat premature as throughout March neither side had managed to gain a significant edge, although the Russian fleet protected by their shore batteries and now under the command of the energetic Admiral Stepan Makarov was perhaps getting the better of the Japanese blockading force under Admiral Tōgō Heihachirō. This changed dramatically on 13 April 1904 when the Russian flagship, *Petropavlovsk,* struck a mine and swiftly sank with the loss of 679 lives including that of Admiral Makarov.[29]

In August 1904, under their new commander Admiral Wilghem Vitgeft, the Russians attempted to break out of Port Arthur with six battleships, four cruisers and fourteen destroyers. Admiral Tōgō, with four battleships, two cruisers and eighteen destroyers, was waiting for them. Initially both sides opened fire at an unprecedented eight miles, and then commenced an indecisive eight-hour sea battle, the Battle of the Yellow Sea, the first ever major clash between steel battleships. The Russians came off worst in the encounter, with Admiral Vitgeft and his entire staff killed when the bridge on his flagship took a direct hit. The battered and demoralised Russian fleet limped back into Port Arthur where they remained until Japanese land forces, advancing from their victories in Korea and Manchuria, encircled the town. The Japanese placed their long range artillery on the hills overlooking the harbour, and the anchored ships trapped within, and systematically sank them one by one like the proverbial fish in a barrel.[30]

[29] The Russians were enthusiastic pioneers in the development of naval mines and had first used them in the Crimean War. The Russo-Japanese conflict was the first time naval mines had been used so extensively by both sides, and proved particularly effective in the narrow confines surrounding Port Arthur. Wells, William Bittle, 1904, *Pacific Monthly, v. XI*, Portland, OR., p. 412.

[30] Schneer, James A., 2015, *Mikasa – An Illustrated History of the Last Pre-dreadnought Battleship*, Morrisville, NC., pp. 25-29.

Meanwhile in one of the most astonishingly ambitious, absurd and ultimately futile naval undertakings in history, Tsar Nicolas II ordered the Baltic Fleet under the command of Admiral Zinovy Rozhestvensky to steam half way round the world to relieve Port Arthur. Unfortunately the poorly trained and inexperienced officers of the fleet, most never having previously served outside the Baltic, were ill-equipped for ocean voyaging. When the Russian fleet on entering the North Sea encountered the British Hull trawler fleet peacefully fishing on the Dogger Bank, it convinced itself that they were Japanese torpedo boats lying in wait.[31] The battleships opened fire indiscriminately on the fishing boats as well as each other with everything they had. One fishing boat was sunk, three British sailors were killed, one by a heart attack, together with a Russian chaplain and sailor caught in the crossfire. It was only the abysmal standard of Russian gunnery that prevented far greater loss of life.[32]

The consequences of this farcical encounter were catastrophic for the Russians. The furious British denied them passage through the Suez Canal and the use of their coaling stations. The fleet was forced around the Cape of Good Hope on an unplanned seven-month journey, with stopovers at Madagascar and French Indochina. It was while at Madagascar that they received the news of the fall of Port Arthur. Because of the enforced detour, the Russian Second Pacific Fleet (as the Baltic Fleet was renamed) was dangerously short of coal and food and was even low on ammunition after pointlessly expending so many shells in the North Sea. Admiral Rozhestvensky made the fateful decision to take the shortest but most predictable route to the shelter and supplies of Vladivostok. Once more Admiral Tōgō and his well trained and well equipped modern fleet were lying in wait. Starting at 13.40pm on 27 May, it took the Japanese just over 20 hours to virtually annihilate the Russian ships in an intense battle fought through the night.[33]

The Battle of Tsushima is one of the great pivotal battles in history. It was the first time a European nation had been defeated by a non-European nation in a modern, full-scale war. It destroyed Russia's power in the Pacific forcing her to sue for a humiliating peace.[34] The devastating loss of prestige suffered by Tsar Nicholas II and his government was a major contributory factor towards the first Russian Revolution of 1905, which itself led directly to the Revolution twelve years later. Their victory had a contrary but equally as calamitous effect on the Japanese. Elated by their achievement in defeating one of the major powers, the Japanese Imperial General Staff were convinced that with more and bigger battleships they could take on and beat any adversary including Britain and America. This conviction would help lead inexorably to the conflicts of the Second World War.[35]

Kisaburo Ohara, a student at Keio University, explicitly acknowledges his debt to Fred W. Rose's Octopus map of 1877.[36] In the text he states, " Black Octopus is a name newly given to Russia by a certain prominent Englishman". [37]

[31] *The Scarborough Evening News*, 24 October 1904 for a contemporary account of the incident. http://www.scarboroughsmaritimeheritage.org.uk/adoggerbank.php

[32] Wood, Walter, 1911, *North Sea Fishers and Fighters*, London, Chapter XXII.

[33] After the epic journey of 18,000 miles (29,000 km) the Russian navy holds the record for the longest voyage ever undertaken for an all-steel, coal-powered battle fleet.

[34] Semenoff, Vladimir, 1907, *The Battle of Tsushima*, London (Second edition translated by Captain A. B. Lindsay).

[35] The tiny principality of Montenegro joined their ally, Russia in declaring war on Japan. Overlooked in the subsequent peace-treaty, Montenegro technically remained at war with the Empire of Japan for 101 years until it eventually signed a separate treaty in 2006.

[36] Harmon, Katharine, 2003, *You are here: Personal Geographies and Other Maps of the Imagination*, Princeton, p. 117.

[37] Baridon, Laurent, 2011, *Un Atlas Imaginaire, Cartes Allégoriques et Satiriques*, Paris, pp. 126-127.

Robida (Albert): L'Appétit de la Pieuvre

Robida (Albert): L'Appétit de la Pieuvre [The Appetite of the Octopus]

coloured lithographed map, (Paris), c.1914, 3 x 5 in. (8 x 13 cm)

This striking French image combines the predatory German Eagle with the grasping tentacles of the octopus to powerful effect. Published just before the First World War, it vividly reflects the French fear of an expansionist and increasingly aggressive Germany.

The artist Albert Robida was a prolific book illustrator and writer of science fiction articles, novels and comics, publishing some 60,000 drawings and 200 books.[38] His work appeared in travel books, magazines, on postcards and other ephemera. He was an autodidact, who left school at the age of 12, and went on to become the founder of *La Caricature* magazine. Like his better known contemporaries, H. G. Wells and Jules Verne, Robida speculated on the form of future warfare. He wrote about and drew prototypes of the tank, armoured cars, bombers, machine guns, electrical communication, submarines, anti-aircraft guns, torpedoes, and even anticipated gas and bacterial warfare.[39] In *La Guerre Infernale*, a novel by Pierre Giffard published in weekly instalments in 1908, Robida's illustrations predicted war between Germany and Britain, the aerial bombing of London, and conflict between Japan and the United States. Disregarded by his contemporaries, and largely forgotten today, Robida's prescience still astonishes.

[38] Emery, Elizabeth, 2011, Albert Robida, Medieval Publicist, in Utz, Richard and Emery, Elizabeth, *Cahier Calin: Makers of the Middle Ages. Essays in Honor of William Calin*, Kalamazoo, MI., pp. 71-75, p. 71.

[39] Speaking of Pictures – A Frenchman foresaw Mechanized War in 1883, *Life Magazine*, 15 June 1942, p. 13.

The London Geographical Institute: Daily Mail War Map of Europe

lithographed map, George Philip & Son, Ltd. London, c.1914, 31 x 22 in. (78 x 54 cm)

This map is interesting for its overview of the size of the combatant forces at the start of the First World War. The stark reminder at how small the British professional army was at the outset of hostilities is a clarion call by *The Daily Mail* for vastly increased recruitment.

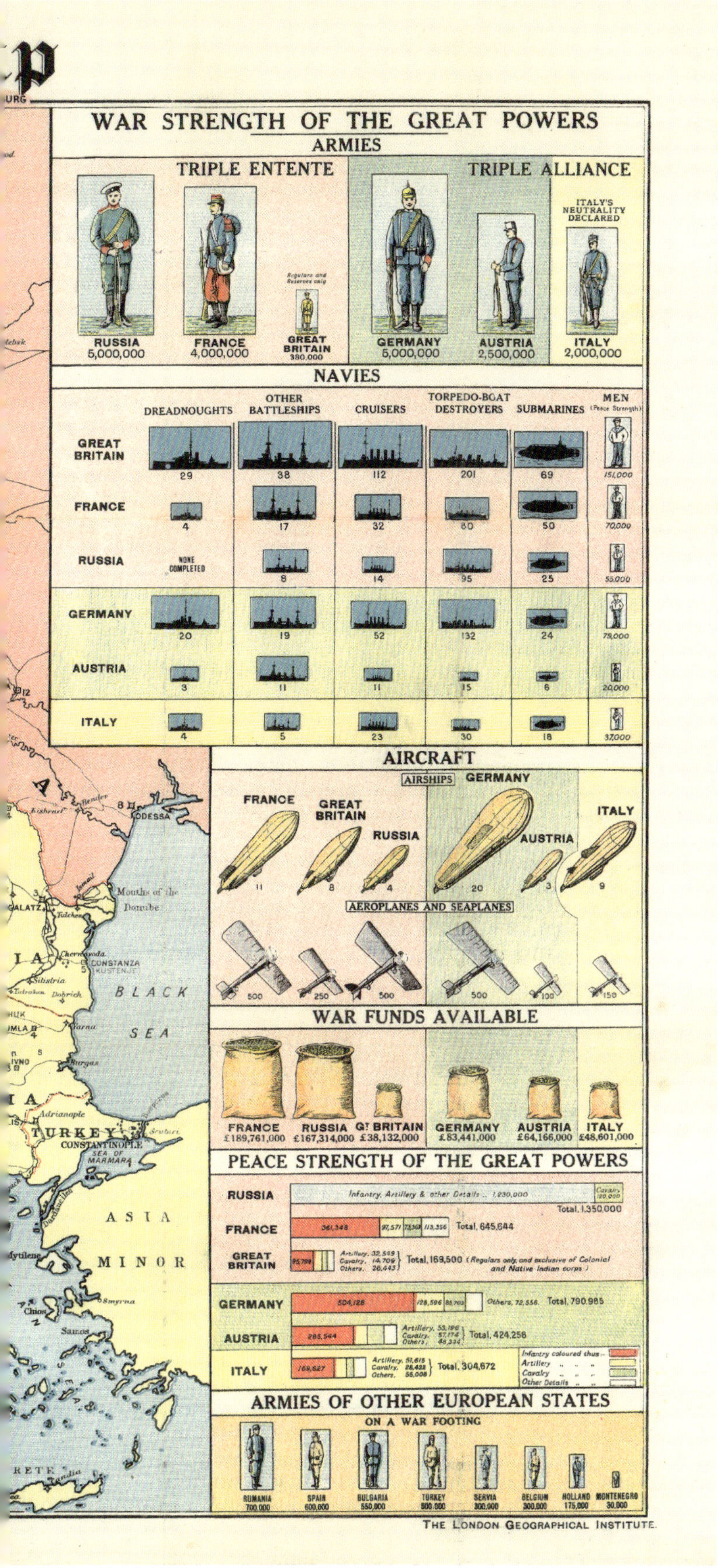

The London Geographical Institute: Daily Mail War Map of Europe

Panels exhibit the strengths of the Commonwealth nations and those of the various nominally unaligned European countries.

The most substantial of these is the Romanian army of 700,000 (actually 650,000), only slightly smaller than that of Britain.[40]

The newest addition to these vast armies was the forces of the air. For the first time aircraft were being used as war machines on a large scale. The greatest danger in the eyes of many military planners was the threat of city-destroying raids by the great airships.[41]

It was in naval power that Britain traditionally had supremacy and this is strikingly demonstrated in the panel dedicated to the navies. Britain has more dreadnoughts, battleships and cruisers than the next two biggest navies combined.[42]

This fascinating summary of the forces at the start of the war would appear to have been printed between the German invasion of Belgium in August 1914 and the First Battle of Ypres in October and November of that year, the first major land engagement of the war and not yet marked on the map.

[40] Keegan, John, 1999, The First World War, pp. 306-308. Although now sometimes overlooked, the Romanian Front in the First World War was seen as of vital significance by both sides who courted the Romanian government, with a bewildering series of secret Byzantine treaties most of which they had no intention of honouring. Eventually Romania joined forces with the Triple Entente in 1916. After some initial successes against the Austrians, the poorly-trained Romanian army collapsed when their Russian allies withdrew their military support after the October Revolution. Forced to sign a truce with Germany in 1917, Romania opportunistically re-entered the war against Germany on 10 November 1918, one day before the Armistice.

[41] The possibility of the aerial destruction of whole fleets and cities was dramatically illustrated in H.G.Wells, *The War in the Air*, published in 1908. Despite the exaggerated terror they inspired, the airships' inaccuracy as bomb-platforms and their vulnerability to attack meant their actual contribution to the war efforts on both sides was negligible.

[42] This was deliberate policy dating back to the early part of the century to ensure British naval superiority, in the event that she should ever find herself at war with both France and Germany at the same time.

Vignal (G.): Il Voudrait l'Univers [He Wants the Universe]

lithographed postcard, published by Fot-Nox, Paris, 1914,
4 x 6in. (9 x 13cm)

This monochrome French postcard mocks the overweening greed and ambition of Kaiser Wilhelm as he attempts to grasp the world. Curiously the German flags, each surmounted by an imperial eagle and liberally scattered across the globe, are depicted with vertical rather than the correct horizontal stripes.

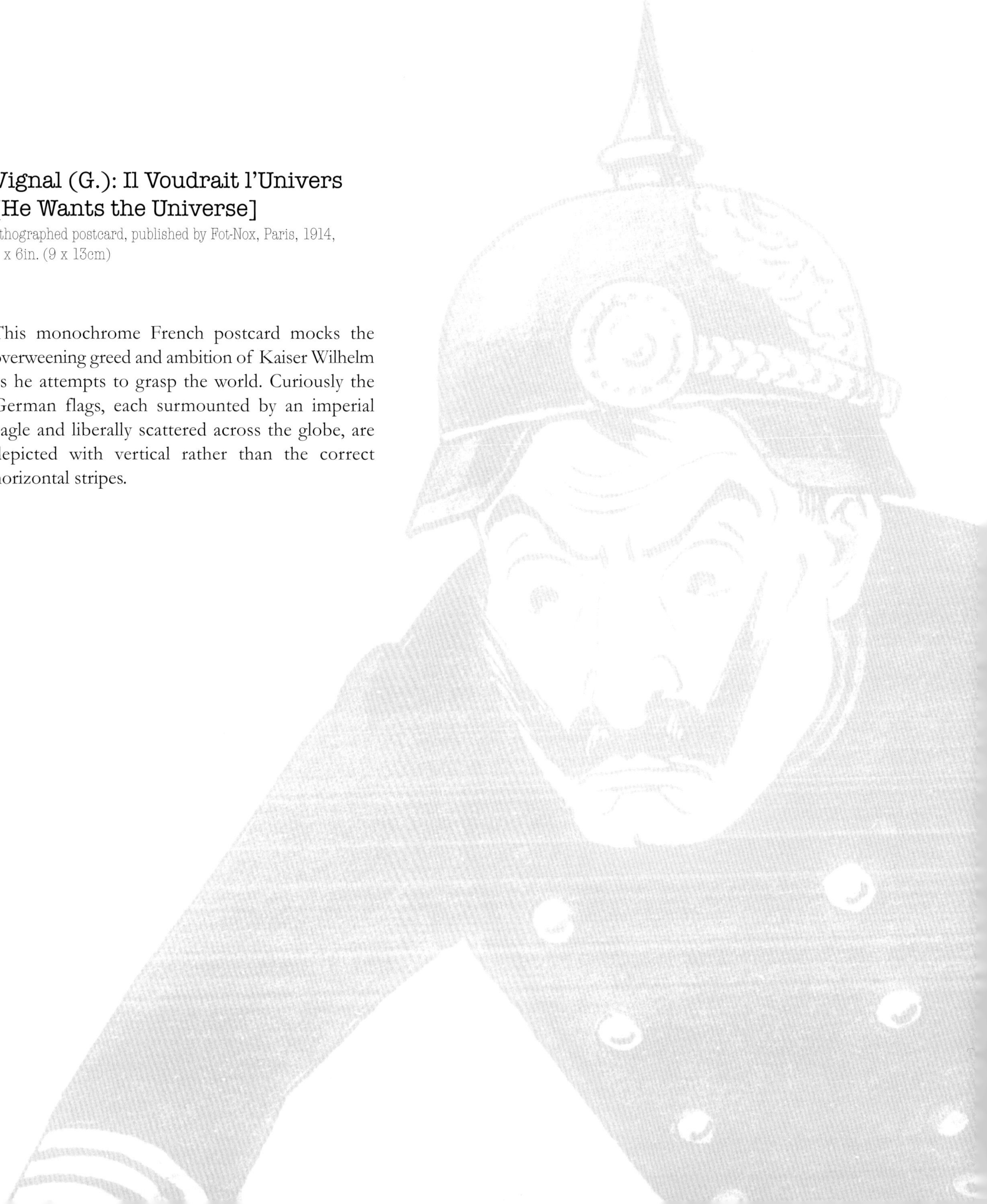

Vignal (G.) Il Voudrait l'Univers

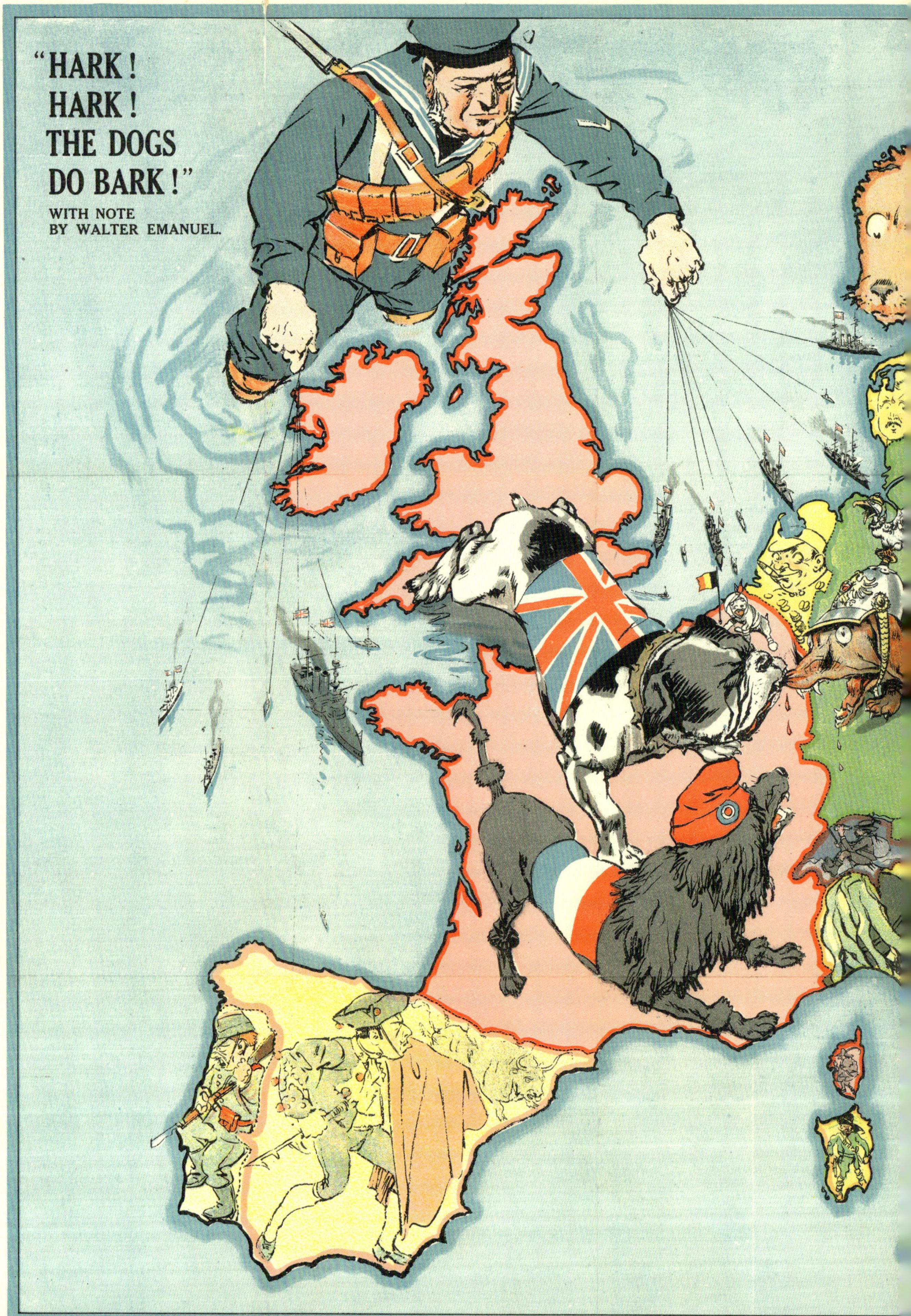

Designed and Printed by Johnson, Riddle & Co., Ltd, London, S.E

The Dogs of War are loose in Europe, and a nice noise they are making! It was started by a Dachshund that is thought to have gone mad—though there was so much method in his madness that this is doubtful. [NOTE FOR THE IGNORANT: The German for Dog is Hund. The English for German is Hun. Dachshund means badger-dog—and he is sometimes more badgered than he likes.] Mated with the Dachshund, for better or for worse, was an Austrian Mongrel. By the fine unwritten law of Dogdom big dogs never attack little ones. There are, however, scallywags in every community, and, egged on by the Dachshund for private ends, the Mongrel started bullying a little Servian. And then the fat was in the fire, for the little Servian had a great big friend in the form of a Russian Bear, and he stood up for his pal. And that was what the Dachshund wanted. He hoped that a big row would ensue, and in the confusion he intended to steal a bone or two that he had had his eye on for some time. He got what he wanted—and a little more. For the Russian B
had friends, too. There was a very game little Belgian Griffon, and there was
great big French Poodle, a smart dandified fellow, and there was a Bulldog. Rat
a sleepy chap this last one, and the Dachshund despised him because he was
always yapping and snarling. But the Bulldog has a habit of sleeping with one
open, and, when he is roused, he grips and won't let go.

The Dachshund started by attacking the Belgian Griffon, as being the small
and mauled the poor creature cruelly, but was quite unable to kill her. And he
mistaken as to the others. He found that the dandified Poodle could fight, and that
Bulldog had not lost the knack of not letting go, and that Russia, after all, was a Rush
and soon the Bear idea made the Dachshund tremble. And even the little Serv
gave the Austrian Mongrel some nasty bites, and so did a neighbour of his named Mo

e & Co., Ltd., London. Published by G. W. Bacon & Co., Ltd., 127, Strand, W.C.

The Dachshund now began to look round for friends, but they seemed strangely arce. He had relied on an Italian Greyhound, a thoroughbred, named Italia, but alia dissembled her love in the strangest way, and asserted that War was a luxury ich she could not afford just now. All the same Italia loaded her gun, and who ows but what it may go off and whom it may hit—for accidents will happen in the st regulated families. The Dachshund, to his annoyance, found only one friend, d that was a dog of Constantinople. The Dogs of Constantinople are quite well own for being fond of offal.

Meanwhile the rest of the European Happy Family looked on, and who shall say w the row will spread? There's the Greek with his knife ready to take a slice of rkey; there are the Balkans determined not to be baulked of their own little ambitions; there's the Spaniard fond of Bull fighting so long as he is not a John Bull; there's the Portugee just spoiling for a scrap; there's the Swiss suffering from cold feet; there's the Dutchman, who keeps smiling with difficulty—still some nice meaty bones may come his way, and in any event he may be relied upon to play the game and not to be a Double Dutchman. [ANOTHER NOTE FOR THE IGNORANT:—Holland used to be known as a low lying country, but this title has now been filched by Germany.] And, up North, the Norwegian, the Swede, and the great Dane all have their eyes well skinned.

All this, and more, may be seen depicted above. Search well and you may find many things. But not Peace. Peace has gone to the Dogs for the present—until a satisfactory muzzle has been found for that Dachshund. Meanwhile the Dachshund's heart bleeds for Belgium—and his nose for Great Britain.

Johnson, Riddle & Co.: Hark! Hark! The Dogs Do Bark!

lithographed map, G. W. Bacon & Co., Ltd., London, 1914, 28 x 21 in. (71 x 53 cm)

This famous image encapsulates the political and military situation at the start of the First World War.

The British Bulldog, supported by a fierce French poodle, is biting the nose of the German dachshund itself tethered to a mongrel, representing the different races of Austro-Hungary.

Behind the canine protagonists we see a determined Tsar Nicolas II driving a steamroller, supported by Cossacks, coming in to flatten the Austrian and German forces from the east, while Bulgaria tries to seize European Turkey.

Britain is dominated by the figure of John Bull wearing a combination of a sailor's uniform and military webbing, and controlling the ships of the Royal Navy on numerous threads.[43] Just above the bulldog's head is a tiny bandaged dog representing battered little Belgium, while a smug Dutchman surrounded by gold coins puffs on a cigar. This image reflects the Allied belief that Holland was unfairly using its neutrality to make a fortune trading with Germany.[44]

The Turkish Empire, Germany's ally, controls the entrance to the Black Sea. The gate is still open, likely dating the map to pre-October 1914, when the Dardanelles were closed to British shipping shortly before Turkey's declaration of war on the Allied Powers.[45]

All this, and more, may be seen depicted above. Search well and you may find many things. But not Peace. Peace has gone to the Dogs for the present—until a satisfactory muzzle has been found for that Dachshund. Meanwhile the Dachshund's heart bleeds for Belgium—and his nose for Great Britain.

Johnson, Riddle & Company, Ltd. was a prominent London printer but unusually is also named as the designer of this map, while George W. Bacon & Co., Ltd., long associated with Frederick W. Rose's celebrated caricatures, is listed as the publisher.

[43] This figure is sometimes described as representing Winston Churchill, then First Lord of the Admiralty but the luxurious side whiskers make it clear that it is in fact John Bull. Furthermore the 'Churchillian' likeness belongs to his appearance at the start of the Second World War rather than the First World War.

[44] In fact the war devastated the Dutch economy, and unrestricted U-boat warfare resulted in the loss of 25 per cent of Holland's mercantile fleet. The mutual hostility engendered by the Allied suspicions was a factor behind Holland giving Kaiser Wilhelm sanctuary at Doorn after the war.

[45] Bryars, Tim and Harper, Tom, 2014, *A History of the 20th Century in 100 Maps*, London, p. 46.

The Dogs of War are loose in Europe, and a nice noise they are making! It was started by a Dachshund that is thought to have gone mad—though there was so much method in his madness that this is doubtful. Mated with the Dachshund, for better or worse, was an Austrian Mongrel. By the fine unwritten law of Dogdom big dogs never attack little ones. There are, however, scallywags in every community, and, egged on by the Dachshund for private ends, the Mongrel started bullying a little Servian. And then the fat was in the fire, for the little Servian had a great big friend in the form of a Russian Bear, and he stood up for his pal. And that was what the Dachshund wanted. He hoped that a big row would ensue, and in the confusion he intended to steal a bone or two that he had his eye on for some time. He got what he wanted—and a little more. For the Russian Bear had friends, too. There was a very game little Belgian Griffon, and there was a great big French Poodle, a smart dandified fellow, and there was a Bulldog. Rather a sleepy chap this last one, and the Dachshund despised him because he was not always yapping and snarling. But the Bulldog has a habit of sleeping with one eye open, and, when he roused, he grips and won't let go.

The Dachshund started by attacking the Belgian Griffon, as being the smallest, and mauled the poor creature cruelly, but was quite unable to kill her. And he was mistaken as to the others. He found that the dandified Poodle could fight, and that the Bulldog had not lost the knack of not letting go, and that Russia, after all, was a Rusher, and soon the Bear idea made the Dachshund tremble. And even the little Servian gave the Austrian Mongrel some nasty bites, and so did a neighbour of his named Monty. The Dachshund now began to look round for friends, but they seemed strangely scarce. He had relied on an Italian Greyhound, a thoroughbred, named Italia, but Italia dissembled her love in the strangest way, and asserted that War was a luxury which she could not afford just now. All the same Italia loaded her gun, and who knows but what it may go off and whom it may hit—for accidents will happen in the best regulated families. The Dachshund, to his annoyance, found only one friend, and that was a dog of Constantinople. The Dogs of Constantinople are quite well known for being fond of offal. Meanwhile the rest of the European Happy Familie looked on, and who shall say how the row will spread? There's the Greek with his knife ready to take a slice of Turkey; there are the Balkans determined not to be baulked of their own little ambitions; there's the Spaniard fond of Bull fighting so long as he is not a John Bull; there's the Portugee just spoiling for a scrap; there is the Swiss suffering from cold feet; there's the Dutchman, who keeps smiling with difficulty — still some nice meaty bones may come his way, and in any event he may be relied upon to play the game and not to be a Double Dutchman. And, up North, the Norwegian, the Swede, and the great Dane all have their eyes well skinned.

All this, and more, may be seen depicted above. Search well and you may find many things. But not Piece. Peace has gone to the Dogs for the present — until a satisfactory muzzle has been found for that Dachshund. Meanwhile the Dachshund's heart bleeds for Belgium — and his nose for Great Britain.

Die Kriegshunde sind in Europa losgelassen und sie machen ein
ging von einem Dachshund aus, den man für tollwütig hielt, trotzdem
heit, daß man daran zweifeln konnte. Verbunden mit dem Dachsh
österreichischer Mischling. Nach dem feinen, ungeschriebenen Gese
größere Hunde kleinere an. Trotzdem gibt es niedere Arten in jeder G
Dachshund für private Zwecke, begann der Mischling den kleinen Serb
Scheckige im Feuer, indem der kleine Serbe einen großen, starken Fre
Bären und dieser sprang seinem kleinen Freunde bei. Und das war
hoffte, daß ein großer Streit entstehen würde und in der Verwirrung be
zu stehlen, auf die er seit einiger Zeit seine Blicke geworfen hatte.
Er bekam was er wollte—und noch ein wenig mehr. Denn der
Da war ein sehr spieliger, kleiner, belgischer Zierhund, und da war ein
ein kluger, stutzerhafter Geselle, und da war eine Bulldogge. Diese le
und der Dachshund verachtete sie, weil sie nicht immer biß und knu
Angewohnheit mit einem offenen Auge zu schlafen und wenn sie
wieder los. Der Dachshund fing an, indem er den belgischen Spielhun
arme Kreatur bös zurichtete, aber nicht imstande war, sie zu töten.
er sich aber. Er fand, daß der stutzerhafte Pudel beißen konnte, und
wohnheit verloren hatte, zuzufassen und nicht wieder loszulassen, und
gänger war, und bald machte der Gedanke an dem Bären den Dachsh
Und sogar der kleine Serbe versetzte dem österreichischen Misc
selbe tat sein Nachbar genannt Monty. Der Dachshund fing jetzt an,

Zimmermann (E.) (Johnson, Riddle & Co.): Horch! Horch! Die Hunde Bellen!

Zimmermann (E.)
(Johnson, Riddle & Co.):
Horch! Horch! Die Hunde Bellen!
[Hark! Hark! The Dogs Do Bark!]

lithographed map, W. Nölting, Hamburg, 1914,
26 x 20 in. (66 x 51 cm)

This is a German version of the previous map. The principal changes are the translations of the title and the text.

The publisher, W. Nölting of Hamburg made something of a speciality of printing caricature maps at the start of the First World War. They produced a number of them, showing the German perspective as well as reproducing virtually unchanged copies of the leading British and French maps.

Occhipint (E.): Dehors les Barbares!!!, [Out with the Barbarians!!!]

chromolithographed postcard, [Paris?], 1914, 3 x 5 in. (8 x 13 cm)

Occhipint (E.): Dehors les Barbares!!!

The figure of Liberty, clasping the flags of Belgium, France, Britain and Russia, is shown crushing the Emperors of Austria-Hungary and Germany beneath a map of Europe.

Gautier (Gilbert): Vous Ne Passerez Pas [You Shall Not Pass]

photo-lithographed postcard, [Paris?], 1914, 3 x 5in. (8 x 13cm)

Gautier (Gilbert): Vous Ne Passerez Pas

This postcard depicts a dignified King Albert of Belgium, wrapped in the Tricolour and Belgian flags blocking the German advance.

Amschewitz (J. H.): Kill That Eagle

Amschewitz (J. H.): Kill That Eagle

lithographed map, Geographia, Ltd., London, 1914,
28 x 19 in. (71 x 48 cm)

Like "*Hark! Hark! The Dogs Do Bark*", this map shows Germany and her allied powers, Austria and Turkey, beset on all sides by the Allies.

At the centre of the conflict are Germany in the shape of an eagle and the Austro-Hungarian Empire as a fallen clown being attacked by the Russian bear and the Serbian terrier. Russia again supported by Cossacks attacks both of the Germanic Empires, while Romania and Bulgaria are on guard with their weapons pointing at Germany and Turkey respectively.

Marianne and *Le Coq Rouge*, symbolising France, are shown defending their nation. Safe in their neutrality, Scandinavia and the Iberian Peninsula are observing the unfolding events in mainland Europe, while little Switzerland cowers in the Alps, covering its ears, waiting for the inferno to subside.

A plainly reluctant Turkey is nervously sitting on a barrel of gunpowder, held in place by the Kaiser.

"*Business as Usual*", inscribed across England, was a term coined in 1914 by Winston Churchill, when First Sea Lord. It referred to the belief that the First World War would not affect Britain's ability to conduct commerce.[46] Representing the forces of the British Empire, soldiers from Canada, Australia, India and Ireland are standing at Britain's side in the North-Atlantic, while the British navy maintains its presence in the North Sea and Eastern Mediterranean.

Italy, indecisive and ignoring her pre-war alliance with Germany, is represented by a tenor singing: *"You made me love you; I didn't want to do it."* First sung by the Broadway star, Al Jolson in 1913, this was one of the best known hits of 1914.[47]

[46] Churchill, Winston, 1981, Speech at the Guildhall, 9 November 1914, *The Oxford Library of Words and Phrases*, Oxford, p. 71.

[47] Al Jolson went on to entertain the G.I.s on the frontlines in both the Second World War and the Korean War. Celebrated during his lifetime as 'the world's greatest entertainer', Jolson famously starred in the first 'talkie', *The Jazz Singer* in 1927. Carringer, Robert. L., 1979, *The Jazz Singer*, Madison, WI.

Amschewitz (J. H.): Europäische Revue – Töte den Adler

Amschewitz (J. H.): Europäische Revue – Töte den Adler [European Revue – Kill That Eagle]

lithographed map, W. Nölting, Hamburg, c.1915,
25 x 20 in. (64 x 51 cm)

Printed shortly after the English edition, the map image remains unchanged in this German version of *Kill That Eagle,* published once more by W. Nölting in Hamburg. Again Nölting has translated the title and added some additional explanatory text under the heading of *"Ein Dokument der Perfiditat Albi*ons*"*.[48] The text subverts the phrase 'business as usual' from a commercial slogan to an overtly military meaning, implying that a state of war is Britain's natural inclination. It is a tribute to the power of the image that both opposing sides are able to invoke it without change, one as a rallying cry against a despised foe, the other as a warning against encirclement by their enemies.

[48] The French first used the term 'Perfidious Albion' in the seventeenth-century to describe Britain, but it came into general journalistic use during the colonial struggles between the two countries in the nineteenth-century. It is interesting to see its adoption here by their mutual enemy, Germany.

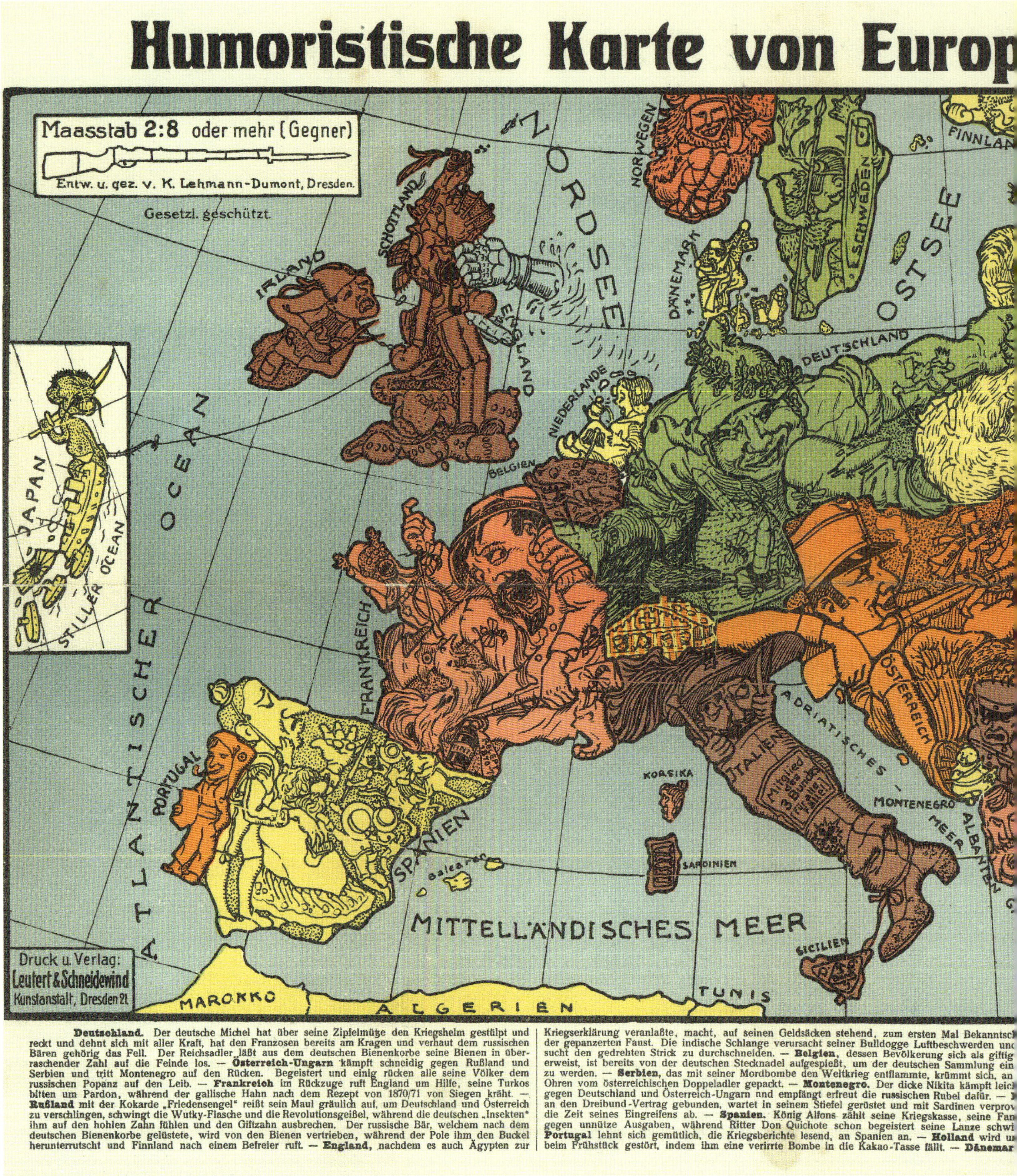

Lehmann-Dumont (Karl): Humoristische Karte von Europa im Jahre 1914

Lehmann-Dumont (Karl):
Humoristische Karte von Europa im Jahre 1914 [Humorous Map of Europe in 1914]

lithographed map, Leutert & Schneidewind Kunstanstalt, Dresden, 1914, 19 x 13 in. (48 x 33 cm)

This is a fascinating depiction of the German view of the situation in Europe in 1914.

A confident German Michel (Der Deutsche Mitchel embodying Germany is the equivalent of Britain's John Bull or the U.S.'s Uncle Sam) is striking out at both France and Russia. The Germans have unleashed a swarm of bees, shown stinging the Russian Bear, the Belgian Toad and the French Cockerel, while the Kaiserliche Marine's High Seas Fleet is a mailed fist emerging from the North Sea to strike England in the face. The German faith in the value of airships for aerial bombardment is demonstrated by the number of zeppelins also shown issuing forth from the beehive, with one attacking England and others en route to Russia. Sultan Mehmed V, representing Turkey, has his back turned on Europe and his arm in a sling, still nursing his wounds from the Balkan Wars of 1912-13. In his left hand is a lit candle impaled on a scimitar, Crimea a puff of smoke, with which he threatens to explode a powder keg under Russia. The scale-bar *(Masstab)* is a pun, referring to the relationship between the Central Powers who, it is claimed, are outnumbered four to one (or more) by the number of their opponents, the Allies.

Meanwhile the Russian giant, drunk and gap toothed, clutches a bottle of vodka in one hand and a whip in the other. Austria-Hungary is focused on the danger from the east while simultaneously stamping on Montenegro and with its Imperial eagle attacking Serbia.

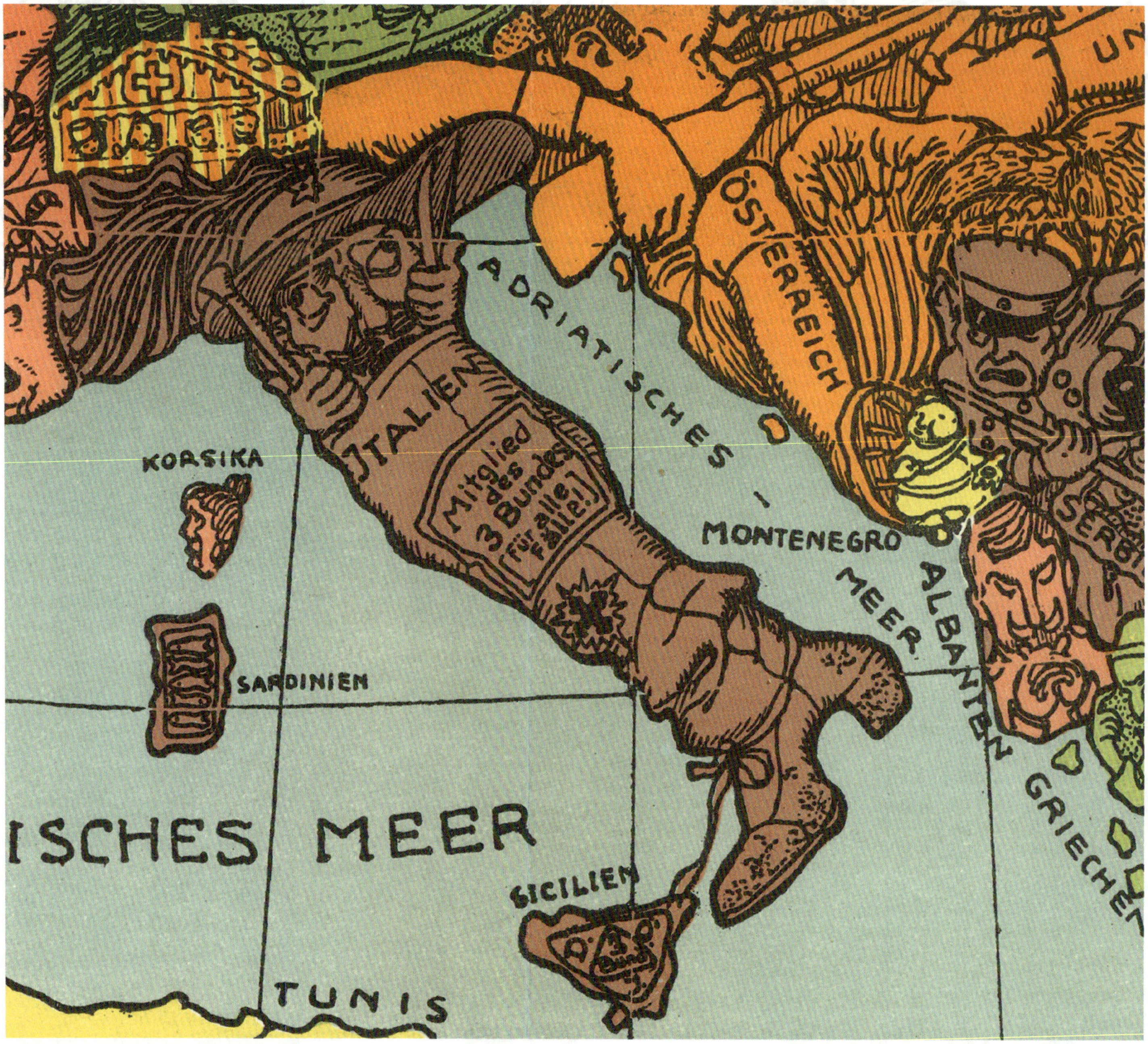

Italy, part of the Triple Alliance since 1882, is peering fearfully from its boot, trying to decide which side to support. They eventually joined the Triple Entente on 23 May, 1915, tempted by the lands in South Tyrol, on the Dalmatic coast, and port cities in Albania, promised by England in the Treaty of London in April 1915.

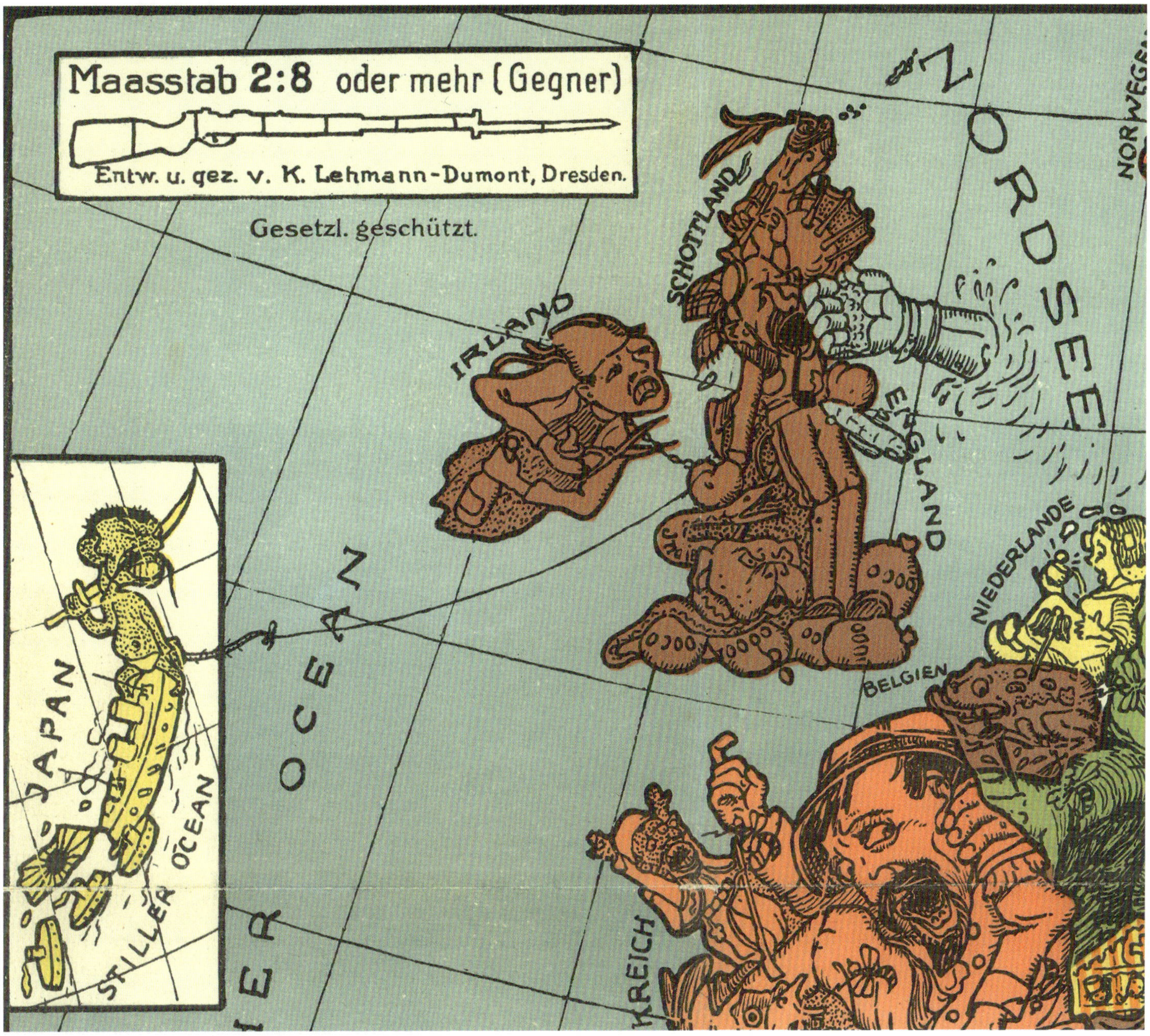

Inset in the far left, a brutish-looking Japan is being dragged into the war by Britain. Ireland, clutching a bottle, tries to cut the chain binding it to England while the Indian Snake attempts to strangle the British Bulldog. This all reinforces the message that cowardly Britain uses her wealth to get other nations to fight her wars.

Other noteworthy details include the bomb in the hand of the Serb, representing anarchy in the Balkans and the assassination of Archduke Ferdinand. In Spain, Don Quixote gallantly attempts to charge into battle, while King Alfonso stands aloof counting his money. The text references both the Danish defeat by Prussia in 1864 and the French loss of Alsace-Lorraine in 1870-71. Belgium is shown as a poisonous toad, reflecting German anger at the Belgium army's dogged resistance to invasion.

It is fascinating to see the very different perspectives of the opposing combatants revealed in these emotive pieces. Germany again is portrayed as the peaceful victim of its neighbours' unprovoked aggression, who with her allies is opposing her attackers' savagery with irresistible force.

2. Humoristische Karte von Europ

Deutschland und Österreich-Ungarn betreiben energisch die Bändigung der im europäischen Tiergarten wild gewordenen, auf sie eindringenden Tiere.

Der **deutsche** Tierbändiger, von zwei Seiten bedrängt, schießt in das auf ihn einstürmende Rudel russischer Wölfe hinein, mit seiner Rechten stößt er kräftig sein Eisen dem französischen Basilisk in den Rachen.

Der **Österreicher** als Tierbändiger schießt ebenfalls auf die Wölfe, treibt mit der Peitsche die freche, blutgefleckte Hyäne von **Serbien** zurück und tritt die bissige **montenegrinische** Ratte unter seinen Fuß.

Der **französische** Hahn, nachdem er das Ei ausgebrütet, dem der gallische Basilisk entkriecht, flattert wild gemacht auf den deutschen Bändiger los, während die **belgische** dreiköpfige Schlange von ihm bereits unschädlich gemacht ist. Ein Schlangenhaupt hängt festgebissen am Ärmel des Bändigers, nur der letzte, dritte Giftrachen öffnet sich noch gegen ihn.

Das **englische** Krokodil lauert mit weit geöffnetem Rachen im Wasser auf günstige Gelege den deutschen Tierbändiger zu verschlingen; die englische Bulldogge hockt sprungbereit und knu im Hinterhalt.

Im Hintergrunde reckt sich auch der **japanische** Drache empor, an seinen Füßen kreisc Affen, und sucht sein feuerspeiendes Haupt gegen den deutschen Bändiger zu richten.

Das große **russische** wilde Rhinozeros stürmt auf die beiden Bändiger ein, vor sich her, im wälzend, ein Rudel Wölfe treibend; den **finnischen** Walfisch hat es fest an der Harpune.

Der **türkische** Elefant sendet, aus der Ruhe geweckt und gereizt, einen kräftigen, kalten W strahl den russischen Bestien entgegen und wird auch den Engländern noch Sorgen bereiten.

Der treue **italienische** Bernhardinerhund ruht auf seinem Lager und hält wachsam die offen.

Lehmann-Dumont (Karl): Humoristische Karte von Europa im Jahre 1914

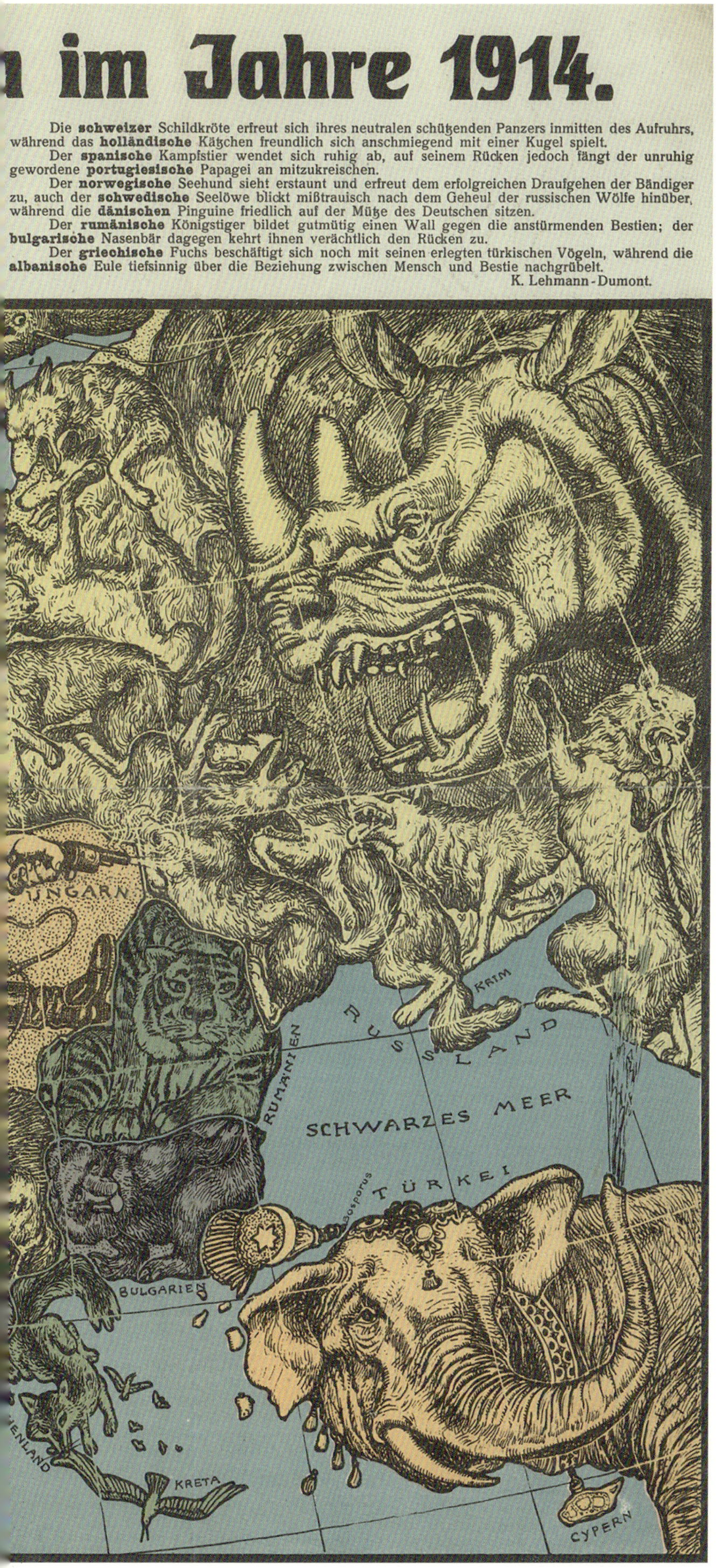

Lehmann-Dumont (Karl): Humoristische Karte von Europa im Jahre 1914 [Humorous Map of Europe in 1914]

lithographed map, Leutert & Schneidewind Kunstanstalt, Dresden, 1914, 19 x 13 in. (48 x 33 cm)

Lehmann-Durmont's second caricature map of Europe takes a further step towards the anthropomorphic, with now all countries except Germany and Austria-Hungary shown as animals. In some cases these are the national symbols or part of the national iconography, as with the bull in Spain, but others are chosen to reflect their perceived threat to the Central Powers.

The map is subtitled *"Deutschland u. Österreich-Ungarn als Tierbändiger"* ("Germany and Austria-Hungary as Animal Tamers") and the two countries are shown defending themselves against a menagerie of marauding beasts. The greatest immediate menace is coming from Russia, and both Germany and Austria are emptying their pistols into the snarling pack of Russian wolves, which are as intent on fighting amongst themselves as getting at their enemies reflecting the civil unrest convulsing Russia. Unusually the country itself is depicted as a monstrously fanged rhinoceros.

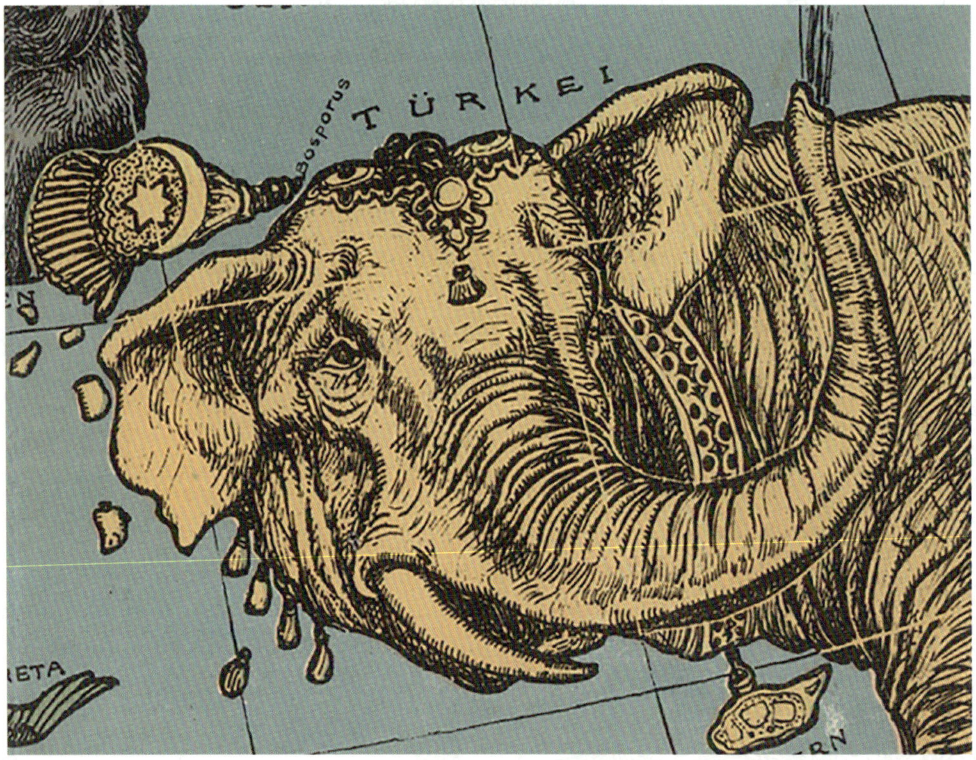

Turkey, the lethargic ally of Germany, is an elephant languidly squirting water from its trunk towards the Caucasus.

Britain is represented by a crocodile whose mouth is partly bound by a strap representing India.[49] The role of the bulldog usually given to England in other caricature maps is here passed on to Ireland.

[49] Germany had high hopes that Indian Nationalists, as led by the likes of Rashbehari Bose, would incite a subcontinent-wide insurrection against their colonial masters which could spread throughout Asia.

France's symbol, the cockerel, is still present but as with the British bulldog, it has been relegated to a subservient position by a more vicious animal, in this case, the basilisk, the mythical 'king of serpents' capable of killing with a single glance. Yet again Belgium still has to bear the German resentment of their impertinent resistance to invasion and is this time depicted as a three-headed snake with two of the heads already severed.

Italy, still trying to make up her mind as to whether to honour her treaty obligations to Germany and Austria, is presented as a dog lying half-asleep. The countries in northern Europe not directly involved in the First World War are symbolised by more peaceful animals: Denmark as penguins, Norway and Sweden as a seal and a sea lion, and Finland as a whale. This is also the case for some countries in southern Europe, such as Albania shown as an owl, or northern Africa in the shape of a camel. Switzerland is a tortoise safe within her shell of neutrality.

The strategic significance of still neutral Romania, with its army of 650,000 men threatening Austria, Germany and Russia, is reflected in her representation as a currently noncommittal tiger.

This map, like Lehmann-Durmont's earlier example, stresses the German and Austrian viewpoint that the cause of the First World War was the rapacious enmity of their neighbours, rather than their own acts of aggression.

Anon: [Wilhelm's Nightmare]

Anon: БРЕДЪ ВИЛЬГЕЛЬМА [Wilhelm's Nightmare]

coloured lithographed map, I. M. Mashistova, Moscow, 1914, 16 x 25 in. (40 x 62 cm)

Т-во Типо-Литографiи И. М. Машистова. Москва

This scarce Russian pictorial piece was published in Moscow, and approved by the Moscow military censors on 23 October 1914. It reflects the belief, prevalent at least among the more optimistic Allied Powers at the start of the First World War, that the huge Russian army would advance inexorably out of the east rolling over the Germans as they came.[50]

The German Emperor, Wilhelm II, his bedspread pulled back to reveal a map of Europe, is shown waking from a nightmare to find a Cossack with levelled lance charging straight into his gaping mouth. Below this striking image is a four stanza poem by R. Mech contrasting Wilhelm's aggressive hubris with the impending destruction of all his dreams.

There had been a long tradition in Russia of popular graphic and satirical prints called 'luboks' or 'lubki'. First appearing in the latter half of the seventeenth-century as inexpensive woodcuts, by the mid-nineteenth-century they were generally lithographed.[51] Originally principally decorative and religious in subject, they became increasingly mocking and subversive over time. This came to a head during Russia's catastrophic and unexpected defeat by the Japanese in the Russo-Japanese War of 1904-5 when the lubok makers took the opportunity to pour derision on the leaders, both political and military, held culpable for the disaster. This resulted in a crackdown on lubki publishers hence the obligatory permission to publish required by the *Moscow Military Censors.*

[50] In 1914 Russia had a standing army of nearly 6,000,000 - 1,500,000 more than Germany, the next largest combatant.

[51] Nedd, Andrew M., Segodniashnii Lubok: *Art, War, and National Identity* in James, Pearl (ed.), 2010, *Picture This: World War I Posters and Visual Culture*, Nebraska, pp. 241-267, p. 241.

J. C.: La France Accueille la Belgique [France Welcomes Belgium]

coloured lithographed postcard, Edition Rojoux & Schaufelberger, Geneva, 1914,
4 x 6 in. (9 x 14 cm)

J. C.: La France Accueille la Belgique

Published in Switzerland, this propaganda postcard was intended for the French market.

"*Remember Poor Little Belgium*" was a rallying cry for the Entente Powers throughout the First World War. Here, Belgium is represented as a vulnerable young girl seeking shelter beneath the French Tricolour.

Ehrmann (E.): L'Attaque et la Riposte [Attack and Response]

lithographed maps (series of six postcards), Edition Lorraine, 1914, each 6 x 4 in. (15 x 10 cm)

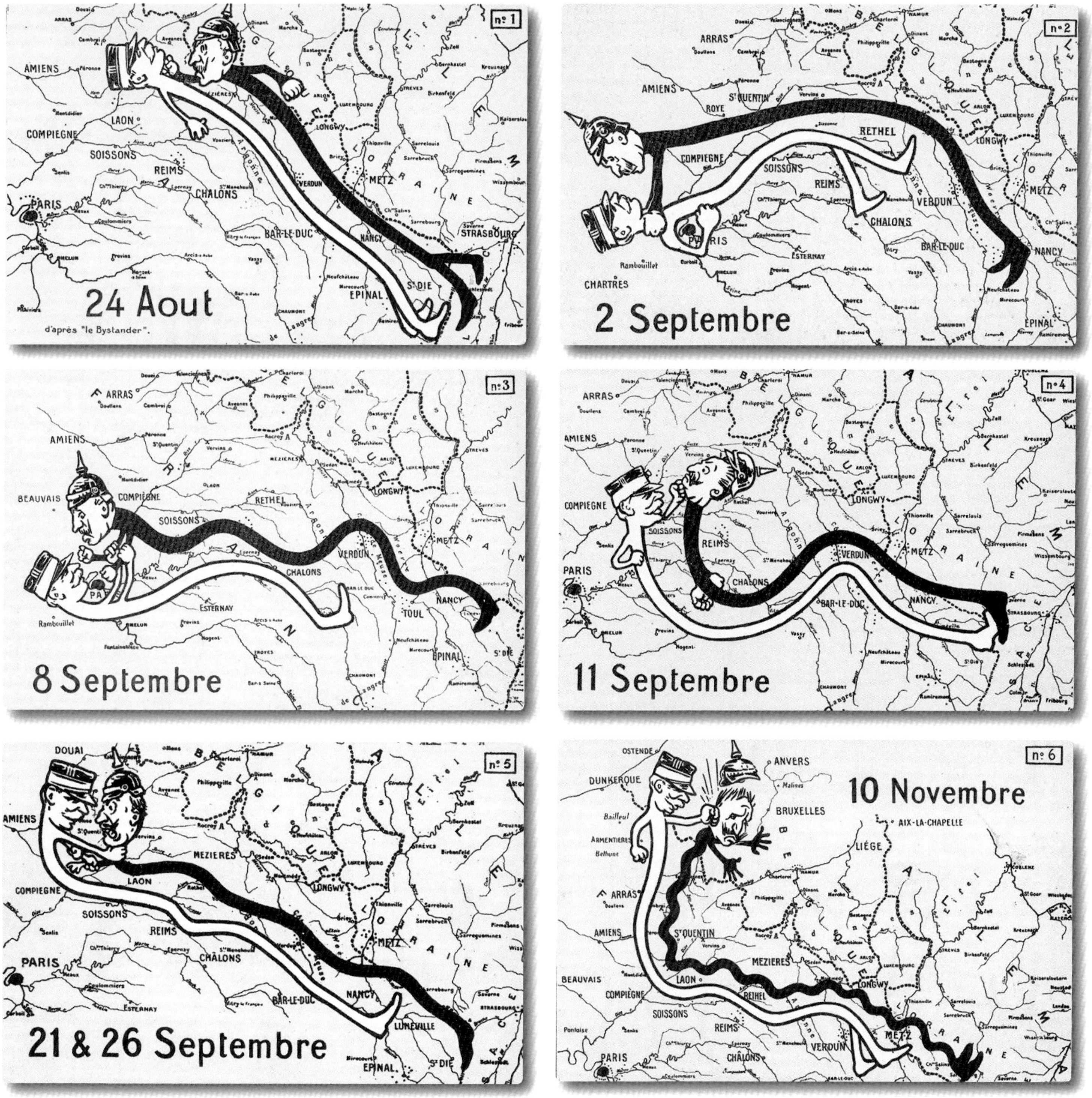

Ehrmann (E.): L'Attaque et la Riposte

Published as a set of six postcards, this series is a striking representation of the movement of the fronts during three months at the beginning of the First World War before they became mired in the stalemate of trench warfare.

Kaiser Wilhelm and Joseph Joffre, the French Commander-in-Chief, represent the opposing sides. Their snakelike bodies mark the situation of the front, while their faces and fists denote the positions of the principal battles.

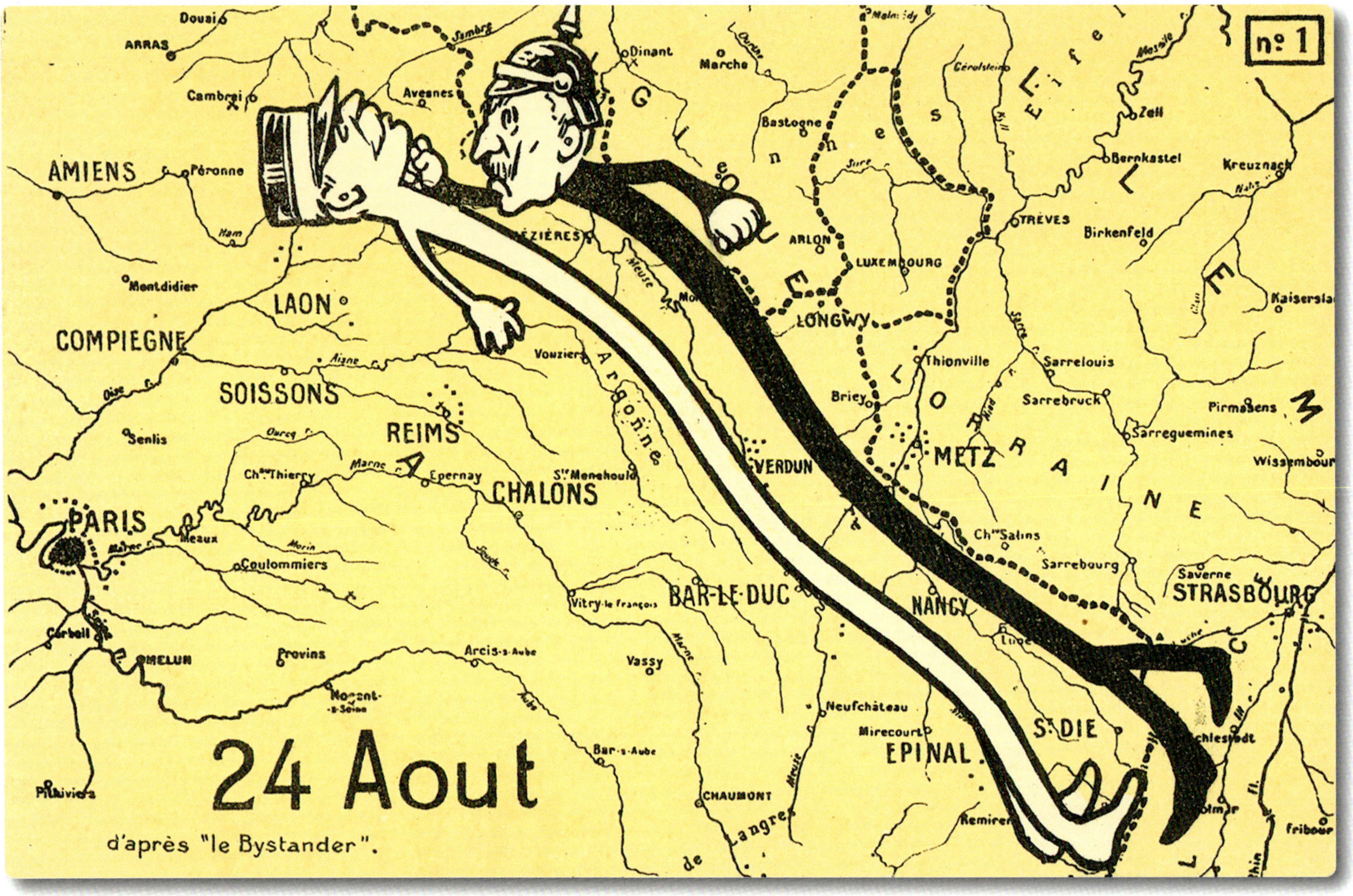

The first map, dated 24 August, shows the Battle of the Frontiers, taking place in France and southern Belgium between 14-24 August 1914. After weeks of tension Germany officially declared war on France on the 3 August, and the complicated interlocking alliances of the Triple Alliance and the Triple Entente swept almost all of Europe into war.[52] The German Schlieffen Plan, drawn up in 1904-05 by Count Alfred von Schlieffen, Chief of the Imperial German General Staff, was designed to swiftly encompass the fall of France before her ally, Russia, could enter the fray and force Germany to fight on two fronts.[53] The first ten days of fighting was a collision between the Schlieffen Plan and the French counter, Plan XVII. Both sides tried to utilise the lessons derived from the successful German campaign of the Franco-Prussian War of 1870-71. Mobility, the use of the railroads and strong reserve forces had been key to Prussian success, and the French were determined to use these factors to go onto the offensive.[54]

Despite their attempts to concentrate their reserves and support from the British Expeditionary Force fighting at Mons, the French continued to be forced back to within 10 miles of Paris in what came to be known as the Great Retreat, with only the stronghold of Verdun continuing to hold the Germans.[55]

The First Battle of the Marne, 5-12 September, marked the end of the Great Retreat by the Western flank of the Allied armies. After a tense struggle, illustrated in the wrestling in map three just outside Paris, the French-Anglo army began to strike back and push the German army away from the capital.[56]

[52] Skinner, H. T. & Stacke, H. Fitz M. 1922, Principal Events 1914-1918, *History of the Great War Based on Official Documents*, London, p. 6.

[53] Humphries, M. O. & Maker, J. 2013, Der Weltkrieg: 1914 The Battle of the Frontiers and Pursuit to the Marne. *Germany's Western Front: Translations from the German Official History of the Great War, I, part 1*, Waterloo, Canada, pp. 66-69.

[54] Strachan, H. 2001, To Arms, *The First World War I*, Oxford, pp. 194-198.

[55] Tyng, Sewell, 1935, *The Campaign of the Marne 1914*, New York, p. 128.

[56] Skinner, H. T. & Stacke, H. Fitz M. 1922.

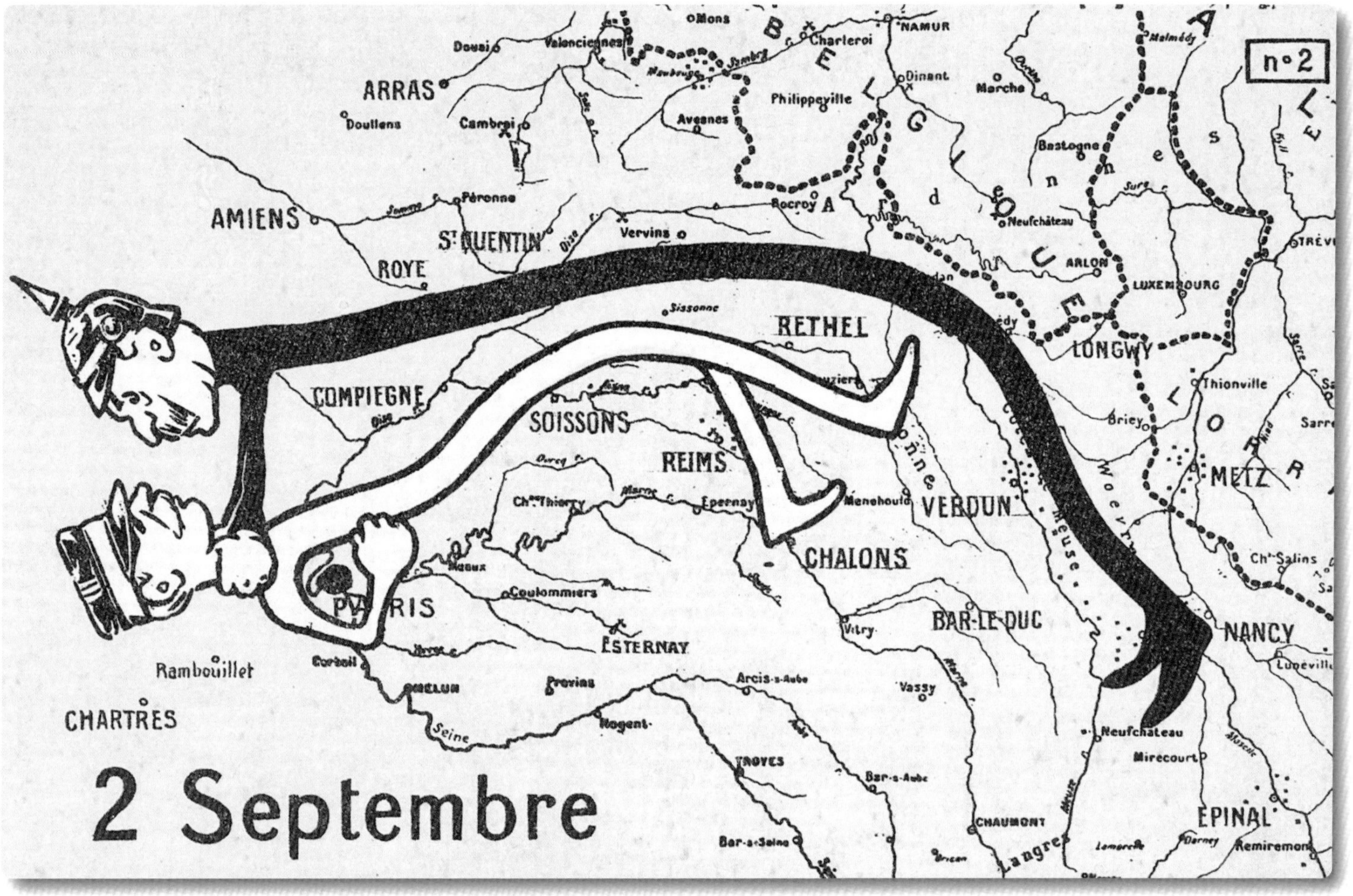
n°2
2 Septembre
ARRAS
AMIENS
St QUENTIN
ROYE
COMPIEGNE
SOISSONS
REIMS
RETHEL
VERDUN
CHALONS
PARIS
BAR-LE-DUC
NANCY
METZ
LONGWY
ESTERNAY
Rambouillet
CHARTRES
EPINAL

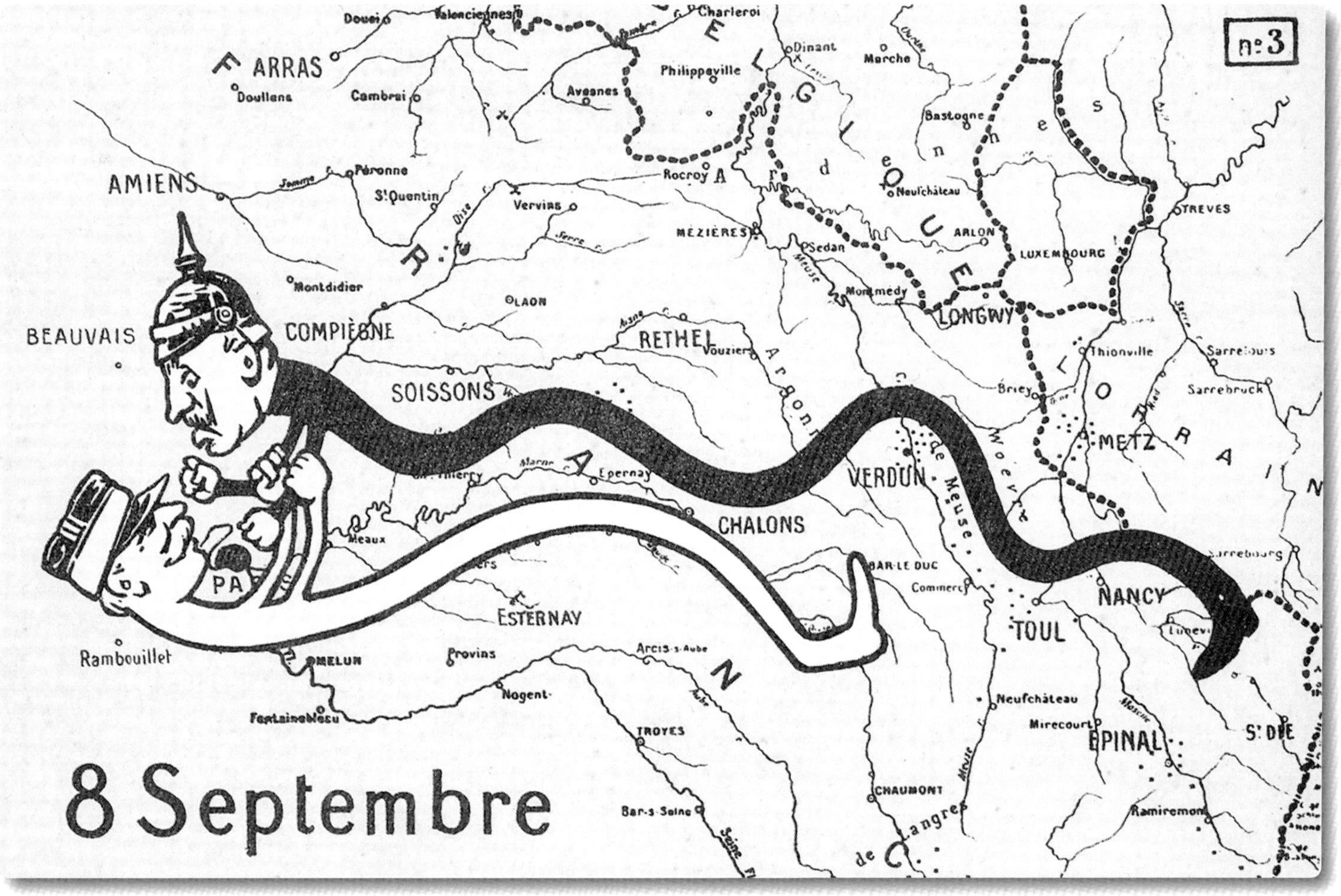
n°3
8 Septembre
ARRAS
AMIENS
BEAUVAIS
COMPIÈGNE
SOISSONS
RETHEL
VERDUN
CHALONS
BAR-LE-DUC
TOUL
NANCY
METZ
LONGWY
ESTERNAY
Rambouillet
EPINAL

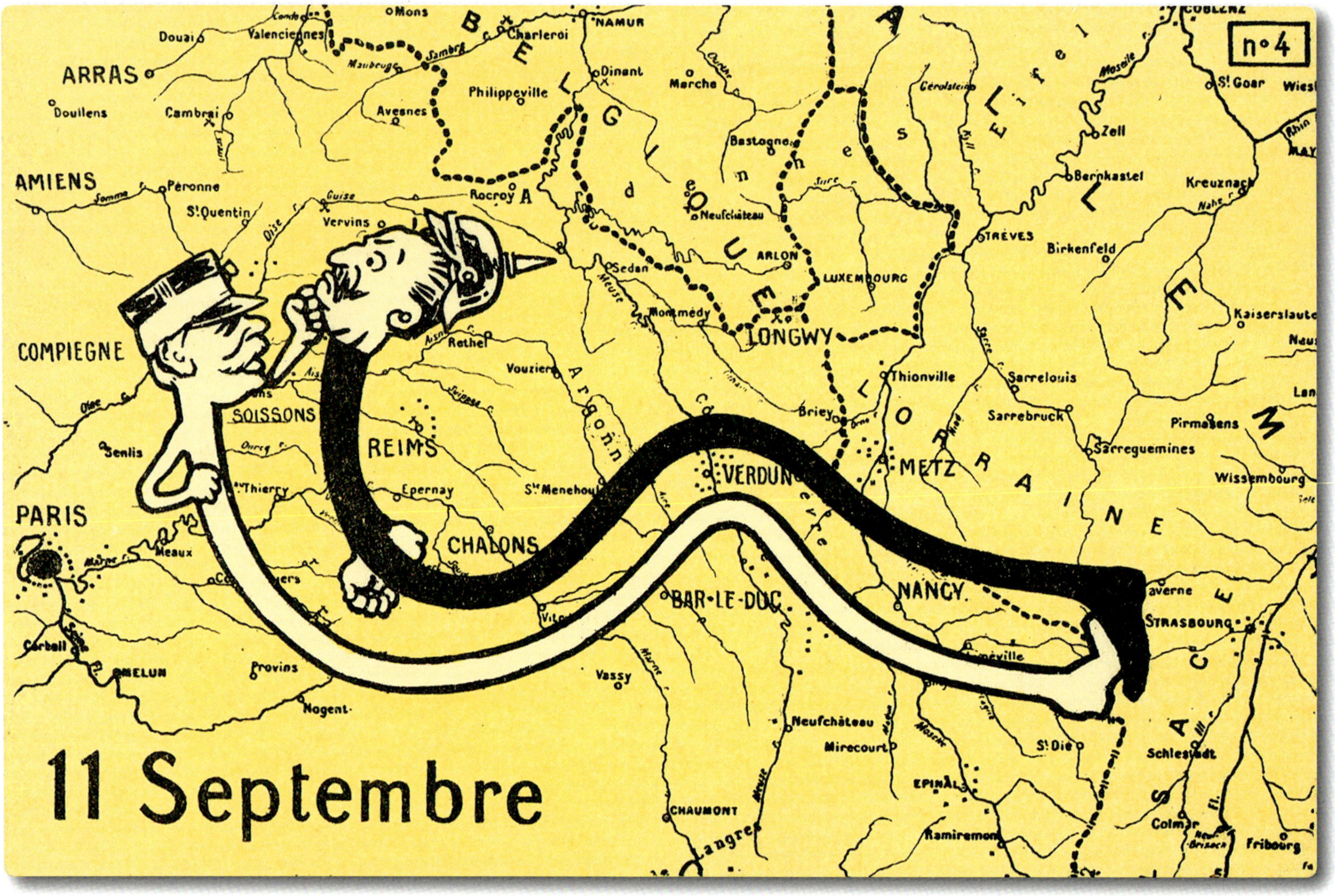

By 10 September the German Army was in partial retreat, and the French pushed them back beyond Verdun once more. To cut off the withdrawal, General Joffre ordered an outflanking move towards the north on 12 September, but it was too slow and the Germans managed to dig in two days later near the river Aisne.

The First Battle of Picardy was fought from the 22-26 September, and the failure of either side to manage a decisive victory against defensive positions was to be typical of the stalemate to come.

The Germans, now immobilised along most of the front, tried to outflank the Allied Powers to the north resulting in a series of battles, in a chain extending to the Channel in what has come to be known as the Race to the Sea.[57]

The final map is dated 10 November, the last day of fighting in The First Battle of Ypres, the culmination of the Race to the Sea. The halting by the British of the attempted German breakthrough signalled the end of the war of movement, and the beginning of four grim years of trench warfare. The frontlines represented in the last two maps would remain virtually unchanged until 1918.[58]

[57] Rickard, J. 2007, *First battle of Picardy, 22-26 September 1914* (http://www.historyofwar.org/articles/battles_picardyI.html)

[58] Sheffield, Gary and Todman, Gary (eds.) 2007, *Command and Control on the Western Front: The British Army's Experience 1914–18*, London, pp. 14, 22, 27, 35, 100.

n° 5
21 & 26 Septembre
DOUAI
AMIENS
COMPIEGNE
SOISSONS
REIMS
PARIS
LAON
MEZIERES
CHÂLONS
BAR-LE-DUC
NANCY
LONGWY
METZ
LUNEVILLE
EPINAL
St DIE
LORRAINE
BELGIQUE

n° 6
10 Novembre
OSTENDE
DUNKERQUE
ANVERS
BRUXELLES
AIX-LA-CHAPELLE
LIÉGE
ARMENTIERES
ARRAS
AMIENS
St QUENTIN
MEZIERES
BEAUVAIS
LAON
COMPIÈGNE
RETHEL
LONGWY
SOISSONS
REIMS
METZ
VERDUN
PARIS
CHÂLONS
LORRAINE
BELGIQUE

Boby (R.): Le Coup de Balai ...[Sweeping Up...]

Barbot & Gallon, Tours, France, 1914, 6 x 4 in. (14 x 9 cm)

A scarce French postcard of a giant French soldier wearing a red cap of Liberty and supported by a British, a Belgian and an Indian soldier, sweeping the Germans out of France after the First Battle of the Marne in September 1914.

Boby (R.): Le Coup de Balai ...

Kleukens (Friedrich Wilhelm): Zeichnet Kriegs-Anleihe für U-Boote gegen England,

Kleukens (Friedrich Wilhelm): Zeichnet Kriegs-Anleihe für U-Boote gegen England, [Purchase War Bonds for Submarines against England]

Mainzer Anzeiger, Mainz, 1914, 8 x 11 in. (20 x 28 cm)

This striking German propaganda image is an appeal for war bond funds to support the growing U-boat fleet.

Despite appearing to encircle the United Kingdom with an unbreakable blockade, the Germans in fact only had twenty-nine U-boats completed at the beginning of the war, of which twenty were ready to go to sea. The shocked British got a taste of what these novel war machines could accomplish when on 22 September 1914, off the Dutch coast, a single U-boat under the command of Lt. Otto Weddigen managed to sink the cruisers HMS *Aboukir*, *Cressy* and *Hogue* in less than an hour, with the loss of nearly 1,400 lives.

At first the U-Boat fleet followed the accepted rules of naval war, leaving civilian and neutral merchant ships alone. But in February 1915 the Kaiser declared the seas around Britain an unrestricted war zone which led to the sinking of the Cunard liner, RMS *Lusitania,* off the coast of Ireland killing 1,198 seamen and passengers, many of them American. The furious American reaction and worldwide condemnation of the attack on an unarmed civilian liner forced the Germans to reinstate the ban on attacking non-military shipping, once more severely limiting the U-boats' effectiveness.[59] After the failure to destroy the British Grand Fleet's command of the sea at the Battle of Jutland, the Germans stepped up submarine anti-commerce raiding in the belief that Britain could be forced into capitulation before the United States could enter the war. Despite the sinking of over 1.4 million tons of shipping between October 1916 and January 1917, the British managed through the increased use of convoys to keep the vital supply lines open. In February 1917 the now desperate Germans once more announced the resumption of unrestricted submarine warfare. On 17 March three American merchant ships were sunk, triggering the U.S. declaration of war against Germany on 6 April 1917.[60]

The artist Friedrich Kleukens studied at the Arts and Crafts School in Berlin, and in 1900 he founded the Steglitzer Workshop for Artistic Prints (Steglitzer Werkstatt für künstlerische Drucksachen) together with F. H. Ehmcke und Georg Belwe. Having taught in both Leipzig and Darmstädt, he later became the director, together with his brother Christian, of the Ernst-Ludwig-Presse. This was an influential arts and crafts publisher founded in Darmstädt in 1907 by the Grand Duke of Hessen[61], yet another of Queen Victoria's grandsons, and the great-uncle of the Duke of Edinburgh, Prince Philip.

[59] Manning, Martin J., Wyatt, Clarence R. (eds.), 2010, *Encyclopedia of Media and Propaganda in Wartime America*, Santa Barbara, CA., pp. 470-473. The Lusitania was in fact carrying quantities of ammunition for the Allies but this was not generally known at the time.

[60] Clark, George B., 2013, *The American Expeditionary Force in World War I: A Statistical History, 1917-1919*, Jefferson, NC., pp. 9-10.

[61] http://kleukens-archiv.de/flash/LP/teil1/index.html

Einst sass John Bull in seinem Fett
jetzt wird er dünn dank U und Z
Ent-Ente-Enten sieht man fliegen
nur zu Onkel Sam's Vergnügen
Die nordischen drei Köni
Deutschland Östreich über alles, – alles and
CULTURE
Marianne steht
Poincaré fleht
French rät
Delcassé dreht
Marianne steht
Ob Königreich ob Republik
es bringt ihm beides die Kolik
U-Yes
Help Help
Spanien lässt mit vielen Freuden
den little John in's Wasser gleuten
Italien auf dem Stühlchen schaukelt
von rosigen Träumen san
Im grossen heissen Afrika – weiss man nicht recht – was da geschah
A.K.
1915

Kampf (Arthur): Gedrängte Frühjahrsübersicht von Europa im Jahre 1915

Kampf (Arthur): Gedrängte Frühjahrsübersicht von Europa im Jahre 1915 [Overview of the Spring in Europe in 1915]

lithographed map, Verlag von Lucas Gräfe, Hamburg, 1915, 28 x 22 in. (71 x 56 cm)

This interesting German perspective on the situation in Europe is the first of two caricature maps created by Arthur Kampf, a fervently nationalist artist who would go on to be revered by the Nazis. The map is titled *Crowded Spring Overview of Europe in 1915*, and depicts the continent at a high point in the war for the Central Powers.

Germany and Austria are portrayed as dignified empresses, perhaps influenced by the depiction of their sixteenth-century counterparts on the sea charts of Battista Agnese, dominating central Europe while their disciplined ranks of troops drive back the forces of France, Britain and Russia.

Spain is shown trying to tip the British out of Gibraltar, while Turkey has padlocked the Dardanelles above a vignette and legend, celebrating the sinking of six Allied battleships in a Turkish minefield at the start of the disastrous Gallipoli campaign. [62]

On the western front the French army is represented by *Marianne*, dressed in Joan of Arc's armour and proclaiming France's cultural mission, sitting astride a mechanical horse which President Raymond Poincare and Minister of War, Theophile Delcasse, together with General Sir John French, are attempting to crank into action.

[62] Broadbent, Harvey, 2005, *Gallipoli: The Fatal Shore*, Camberwell, pp. 33-34.

On the eastern front a joint German Austro-Hungarian army stands firm against a Russian giant, depicted as a peasant with the ubiquitous vodka bottle in his pocket, having his right hand severed by the 'faithful' Paul von Hindenburg. Russia is described as a fallen colossus and bearer of 'great ideas', probably referring to the pan-slavism which threatened to destroy the Habsburg Monarchy.

Yet again Britain in the guise of John Bull is shown crouched on a bulging money sack, and waving the flags of three neutral countries, Denmark, Holland and the United States, while encircled by U-boats. John Bull is said to have lost weight as a result of the U-boat blockade and the zeppelin attacks while ducks ('Ente' – a play on 'Entente') fly towards Uncle Sam.

Druck von Gebrüder Lüdeking, Hamburg 30

Kampf (Arthur): Sommerschau über Europa 1915

Kampf (Arthur): Sommerschau über Europa 1915 [Summer in Europe in 1915]

lithographed map, Verlag von Lucas Gräfe, Hamburg, 1915, 28 x 22 in. (71 x 56 cm)

This is the companion piece to the previous map, here showing the German view of the situation in Europe in the summer of 1915.

France as Marianne, now on crutches with her toy horse upturned in the sea, is being supported by a British tommy. The vacillating Italy of the earlier map is now a uniformed soldier in the distinctive hat and uniform of the Bersaglieri, and is being struck in the face with an Austrian rifle butt. In the east the Russian giant, now with a bandaged arm, is being bundled back into Russia.

Germany's ally Turkey is shown gorging on Allied warships while the text remarks on everyone's astonishment at the appetite of the 'Sick Man of Europe'.

John Bull is still sitting on his money bags, but is now futilely attempting to fish for further allies using his cash as bait.

The artist Arthur Kampf was born in Aachen in 1864, and studied under Peter Janssen at the Kunstakademie Düsseldorf from 1879 to 1888. He later moved to Berlin and there succeeded Anton von Werner both as Germany's leading historical painter and as director of the Hochschule für Bildende Künste from 1915 to 1924.[63] He joined the Nazi Party in 1933 shortly after they came to power and his work was much admired by Hitler. He was well represented at the *Große Deutsche Kunstausstellung* (The Great German Art Exhibition) held simultaneously as a counterpoint to the *The Degenerate Art Exhibition* in Munich in 1939. In 1944 Kampf was one of 1,041 artists, architects, composers and actors, chosen by Hitler and Goebels and placed on the *Gottbegnadeten List* (God-gifted List), considered crucial to Nazi culture and exempted from military involvement in the final stages of the war.[64]

[63] Lenman, Robin, 1997, *Artists and Society in Germany, 1850-1914*, Manchester, pp. 80-81.

[64] Klee, Ernest, 2007, *Das Kulturlexikon zum Dritten Reich. Wer war was vor und nach 1945*, Frankfurt am Main, p. 294.

Anon: Injustice and Righteousness – [The New Quadruple Alliance and its Enemies]

lithographed atlas, [Berlin?], 1915, 16 x 8 in. (44 x 22 cm)[65]

Anon: [Injustice and Righteousness – The New Quadruple Alliance and its Enemies. Plate 1]

The powerful series of images with text in both Arabic and Farsi in this rare German propaganda booklet were designed to encourage insurrection in the disaffected colonies of the Allied Powers.

Anon: [Injustice and Righteousness – The New Quadruple Alliance and its Enemies. Plate 2]

[65] The 'Quadruple Alliance' was formed in 1915 after Bulgaria joined the Central Powers of Germany, Austro-Hungary and Turkey.

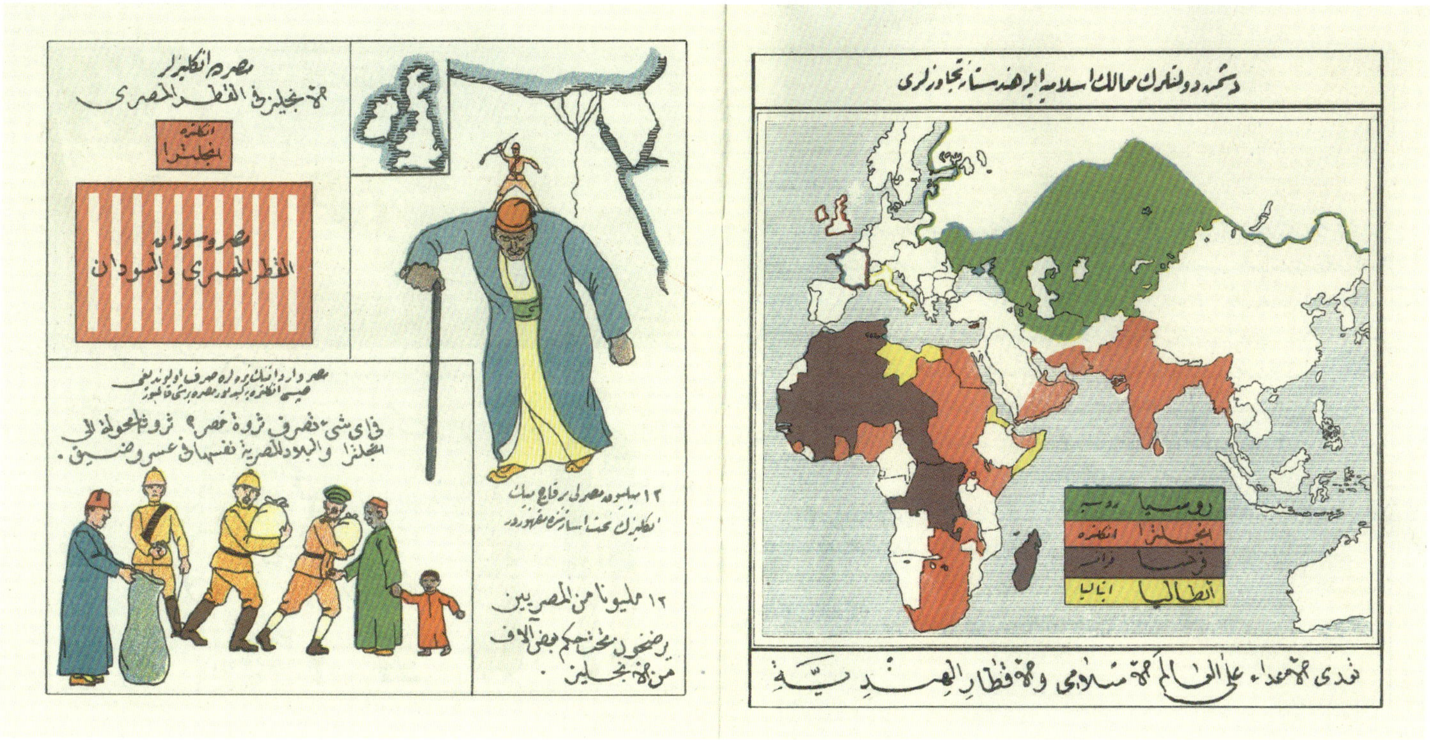

Anon: [Injustice and Righteousness – The New Quadruple Alliance and its Enemies. Plate 3]

It was during the Kaiser's first visit to Istanbul in 1898 that he began to develop his grandiose dreams of a holy war against his enemies. He and his advisors harboured hopes that with the encouragement of their Turkish allies and paternal German guidance, the Muslim peoples of the Middle East would rise up and overthrow their colonial masters. In October of that year he wrote to his cousin, Tsar Nicolas II, that he should never forget that "the Mahometans [are] a tremendous card to play in our game in case you or I were confronted with a war with a certain meddlesome Power [Britain]".[66]

The text and images accompanying the maps, repeatedly contrast the vast size of the downtrodden colonial masses relative to the tiny number of their oppressors.

[66] Röel, John, 2014, *Kaiser Wilhelm II: A Concise Life*, Cambridge, p. 77.

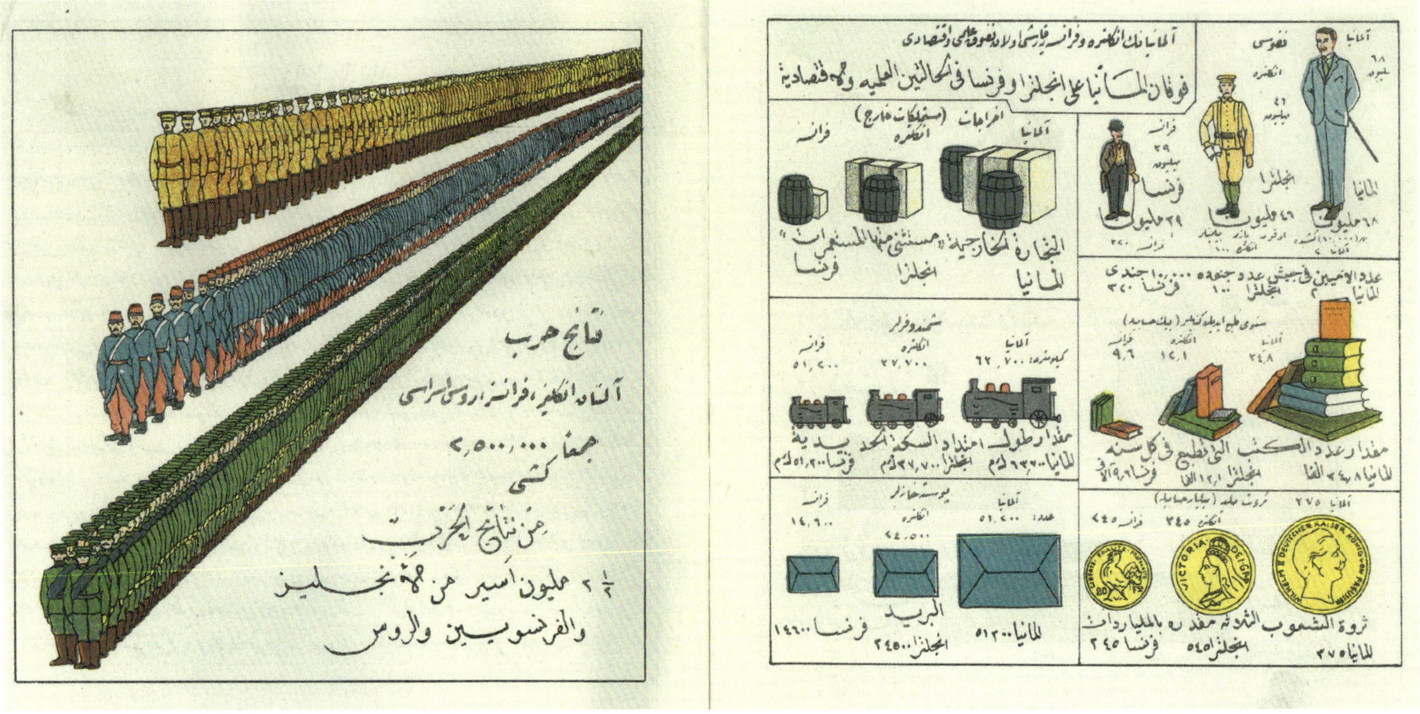

Anon: [Injustice and Righteousness – The New Quadruple Alliance and its Enemies. Plate 4]

An interesting series of graphics compares the relative strengths of Britain, France and Germany in various fields. Germany comfortably comes top in books printed, literacy, population,[67] wealth (excluding colonies), length of railways and the amount of post handled.[68]

India was a principal target for German anti-British propaganda as the population was still reeling after a series of devastating famines, which were blamed with some justification on the colonial rulers. It was the British fear and German hope that the Indian nationalist movements, spurred on by religious fervour, would seize their chance to eject their masters.

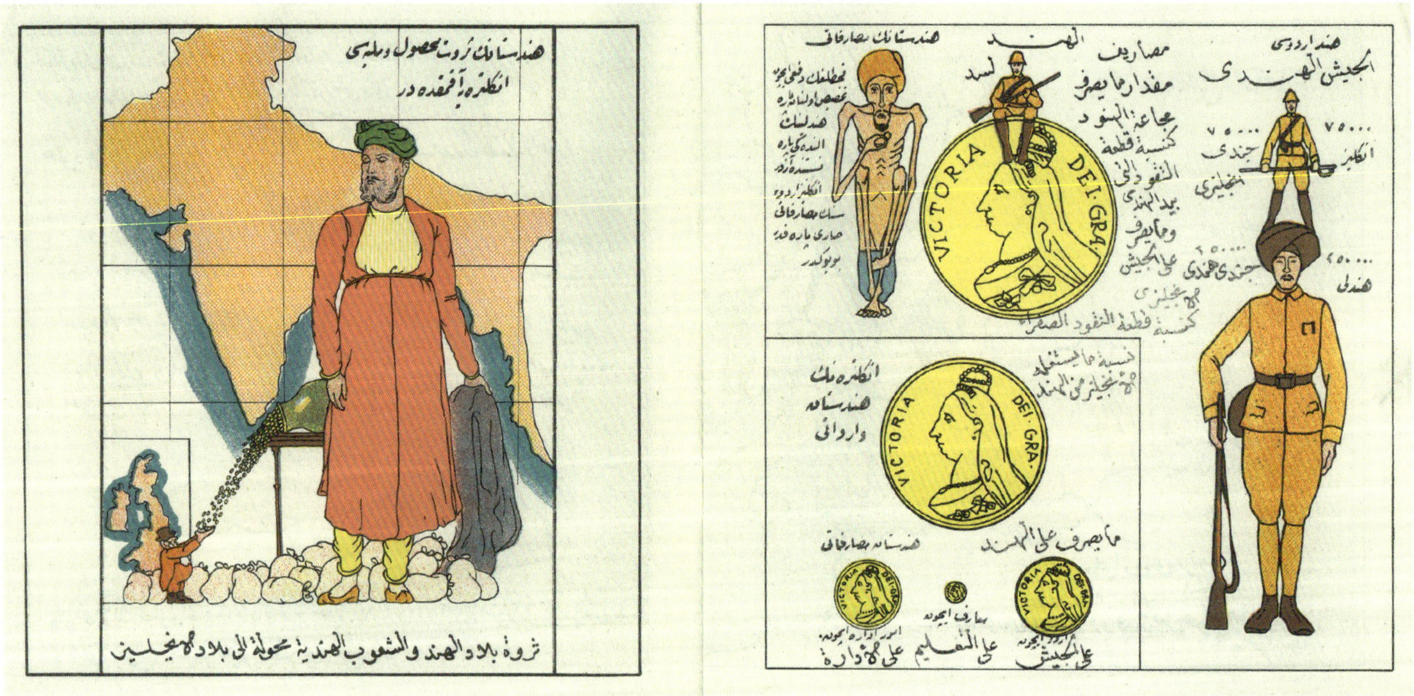

Anon: [Injustice and Righteousness – The New Quadruple Alliance and its Enemies. Plate 5]

The images relating to India focus on the rapacity of the British draining gold from their wretched colony. The most potent image shows a small soldier sitting on a huge gold coin alongside a painfully emaciated giant figure clutching a tiny coin, the coins representing the proportionate amounts spent by the British on the army and on famine relief.

Also targeted are the French colonies of North Africa. An arresting illustration shows a lounging French officer complacently watching a starving farmer ploughing a field with a donkey and camel yoked awkwardly together. Alongside the map are the numbers of the rulers and the ruled – 60,000 Frenchmen in total controlling a population of 5.5 million in Algeria, 3.5 million in Marrakech and two million in Tunis.

Anon: [Injustice and Righteousness – The New Quadruple Alliance and its Enemies. Plate 6]

The final image represents 'the foreign peoples under the cruel enslavement of the tyrannical Russians'.

Wilhelm II's vision of a worldwide colonial uprising against his enemies was destined to be disappointed.[69] On 14 November 1914 Germany's ally, Mehmed V, Caliph and Sultan of Turkey did indeed declare the very first jihad in history against a European nation, but this was not regarded with any enthusiasm outside the Ottoman heartland.[70] The majority of Arabs seized their chance to throw off the oppressive yoke of their detested overlords and allied themselves with Turkey's enemies in what became the 'Arab Revolt', lasting from 1916 to 1918.[71] In India, many of the people rallied to the support of the British Empire with varying degrees of enthusiasm, and 800,000 Indian volunteers fought for the Allies in virtually every theatre of the war. [72]

Anon: [Injustice and Righteousness – The New Quadruple Alliance and its Enemies. Plate 7]

[67] The Germany and British populations are reasonably accurately given as 68 and 46 million respectively but France's population is reduced from 40 million to 29 million, presumably to minimise the unfavourable aggregate total in relation to Germany.

[68] The amount of post handled may seem like an unusual subject for propaganda, but for peoples at the beginning of the twentieth-century for whom the appeal of recent wonders of communication such as radio and the telegraph were limited, this was a potent symbol of modernity.

[69] The idea was not so far-fetched that it wasn't recognised as a genuine threat back in England. In 1916 the novelist, John Buchan, who worked for the War Propaganda Bureau in the First World War, used the idea of a German inspired Islamic uprising against the British Empire as the central plotline in his thriller *Greenmantle*, the sequel to *The Thirty-Nine Steps*.

[70] Marchand, Susanne E., 2009, *German Orientalism in the Age of Empire: Religion, Race and Scholarship*, Cambridge, p. 439.

[71] See: Schneider, James, 2011, *Guerrilla Leader: T. E. Lawrence and the Arab Revolt*, New York; Wilson, Jeremy, 1990, *Lawrence of Arabia*, New York.

[72] See: Pati, Budheswar, 1991, *India and the First World War*, New Delhi.

L'ENTENTE CORDIALE ♥
1915

Von Baumgarten (Eugen): L'Entente Cordiale, 1915

lithographed map, Munich, 1915, 14½ x 17 in. (37 x 44 cm)

This is another strong image reinforcing the German propaganda message that Great Britain's aggressive imperial ambitions were the principal cause of the war, and that she was dragging her misguided allies into the conflict to fight on her behalf.

A monstrous spider with the face of John Bull squats malevolently across Europe devouring a French soldier at Calais. Its web extends to the edge of the territories controlled by the Central Powers. In both the Atlantic and the Black Sea the web is being cut by German U-boats and one leg has already been severed. Further threads entangle the United States, Australia and India. The Dutch and Norwegian flags are flying from the spider's legs. Although neutral countries in the First World War, they were trading partners of the Triple Entente nations, and regarded by the Germans as puppets of the British. A crowned eagle representing Germany looks down imperiously on the scene below.

Eugen von Baumgarten was a prolific German artist, illustrator and cartoonist born in Regensburg in 1865. The artist's monogram, a tree behind a fence, a play on his surname Baumgarten meaning orchard, can be seen in the lower right-hand corner. He was well connected in newspaper and publishing circles in Munich through his cousin, the leading journalist Hermann Roth.[73] Through his friendship with the editor Julius Beck he also illustrated magazines such as *Münchener Humoristischen Blätter*, *Der Bayrische Kladderadatsch*, *Kikeriki-Kalender (Tageblatt)*, *Geißel* and *Jugend*. He designed posters, advertisements and postcards for Münchener Brauhäuser and the Oktoberfest-Zeitung, and ran his own publishing house.[74] Between 1916 and 1918 he made political cartoons and propaganda pieces under the pen name E. Vaube. He died in Munich in 1919.

[73] Roth, Eugen, 1957, *100 Jahre Humor in der deutschen Kunst*, München.
[74] www.lambiek.net/artists/v/von_baumgarten_eugen.htm

Huguenin (André): Théatre de la Guerre Européeanne

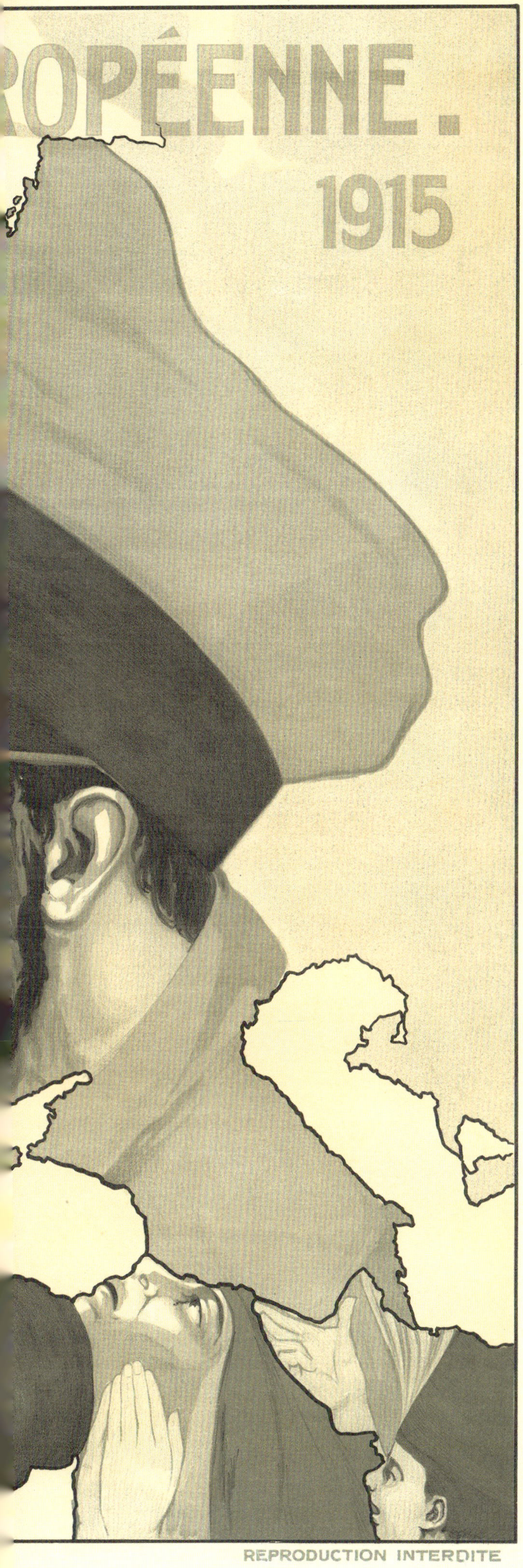

Huguenin (André):
Théatre de la Guerre Européenne
[European Theatre of War]

lithographed map, Aug. G. Fiedler, La Chaux-de-Fonds, Switzerland, 1915, 25 x 22 in. (64 x 55 cm)

This elegant monochrome caricature of Europe, published in Switzerland for the French market, shows the countries characterised by their heads of state or national stereotypes. To the right we see the dominant profile of Nicholas II of Russia, his lighted cigar tip resting on the head of Franz Josef I of Austro-Hungary who in turn is attempting to swallow the Balkans.

Kaiser Wilhelm II is eyeball-to-eyeball with a defiant French 'poilu' and George V looks on holding a bridled fish, representing the Royal Navy's mastery of the sea.[75]

[75] 'Poilu', literally 'bearded one', was an affectionate term for French soldiers going back to Napoleonic times, when they typically sported extravagant beards and moustaches.

Victor Emmanuel III of Italy ineffectively pokes Franz Joseph's ear with a stick while King Peter of Serbia plunges a dagger into his mouth.[76] A vacillating Ferdinand I, King of Bulgaria, turns his view towards European Turkey as Romania, represented by a ferocious sturgeon, heads towards Serbia.[77] Scandinavia, usually shown as a pair of dogs following the example, is here symbolized by a single rabbit. Spain and Portugal, both still neutral at this point, are represented by Alphonso XIII of Spain and Manuel de Arriaga, the first President of Portugal.[78]

The artist André Huguenin-Dumittan spent most of his life, from his birth in 1888 to his death in 1975, in the town of La Chaux-de-Fonds in Switzerland. He attended the Ecole Nationale des Beaux Arts in Paris from 1911 to 1916, and as a sculptor and illustrator was an influential member of several Swiss committees and artistic movements, as well as being closely associated with the famous Swiss artist, Charles L'Eplattenier.[79] Upon his death, the Fondation Huguenin-Dumittan was created by his family to preserve the memory of his work and to promote sculpture by giving an annual award in his name.

[76] Often overlooked as a combatant, Serbia suffered proportionately worse than all the other nations involved in the First World War. Depending on the differing sources, out of a population of 4,500,000 at the start of the war, between 700,000 and 1,200,000 people both military and civilian, had perished by the end.

[77] Despite being related to both Kaiser Wilhelm and Emperor Franz Joseph, Ferdinand was persuaded in 1916 by his wife, Marie of Edinburgh, granddaughter of Queen Victoria, to ally Romania with the Entente Powers. Famed for her beauty and romantically generous disposition, Marie inspired the celebrated lines of Dorothy Parker:

"Oh, life is a glorious cycle of song,
A medley of extemporanea;
And love is a thing that can never go wrong,
And I am Marie of Roumania."

In 1919, Queen Marie personally represented Romania during the negotiations at the Paris Peace Conference to ratify the Treaty of Versailles. The combination of her charm and the gratitude of the other victorious nations achieved enormous political and territorial gains for her adopted country.

[78] After escalating clashes between German and Portuguese colonial forces in West Africa, Germany eventually declared war on Portugal on 9 March 1916.

[79] Collectif de Recherches de l'Université et Musées Lausannois, 1986, *19-39: la Suisse Romande entre les Deux Guerres*, Michigan, p. 61.

Anon: F.. tons l'camp ! Il fait trop de vent ici!! [Let's get out! It's too windy here!!]

lithographed postcard, Paris, c.1915, 6 x 4 in. (15 x 10 cm)

King George V of England and Raymond Poincaré, President of France, hands clasped in amity across the Channel, are blowing Kaiser Wilhelm II back out to sea. Somewhat incongruously, considering the lack of Russian naval activity outside the Baltic and Black Seas during the First World War, Tsar Nicholas wearing a Cossack uniform is pictured in a gun boat in the North Sea. In a cruel jibe, the Kaiser's disabled left arm is shown strapped to his body by his belt.

Anon: F.. tons l'camp ! Il fait trop de vent ici!!

Tiefdruck von Knackstedt & Co., Hamburg 22.

Zensur Altona No. 831

Eddelbüttel (Richard): Europa im zweiten Kriegsjahr 1915/16

Eddelbüttel (Richard): Europa im zweiten Kriegsjahr 1915/16 [Europe in the Second Year of the War 1915/16]

lithographed map, Knackstedt & Co. for Hindenburg-Cognac-Vertriebs-Gesellschaft m.b.H, Hamburg, 1916, 23 x 16 in. (58 x 40 cm)

This entertaining satirical map of war-torn Europe was created to use German patriotism to encourage the sale of Hindenburg brandies and liqueurs.

The text in the title cartouche reads: *"Germans! Demand German products! German brandy. German liqueur. [Get] Hindenburg of the same quality and vintages as supplied to the Officers' mess at the main Headquarters of His Majesty Kaiser Wilhelm II."* It goes on to state that *"Special permission to use the name 'Hindenburg' was granted by His Excellency Field Marshal General von Hindenburg."*

The British lion is being menaced by the German eagle, while a fearful John Bull, crouched as usual over his money-bags, has only a battered umbrella to protect himself from air attacks. A British spider extends its long legs across to France where a haggard Marianne averts her eyes from the front. President Poincaré pours gold into the patched boot of Italy while the Austrians simultaneously pour in shells.

In Spain a flamenco dancer sprinkles pepper on a British bulldog hiding under her skirts in Gibraltar.

By 1916 the war on the eastern front had reached a stalemate similar to that in the west which continued until the Russian collapse in 1917. The front line stretched from Riga, through Jakobstadt (Jēkabpils) and Dünaburg (Duagavpils) in modern day Latvia, Baranavichy and Pinsk in Belarus, to Dubno and Ternopil in western Ukraine.

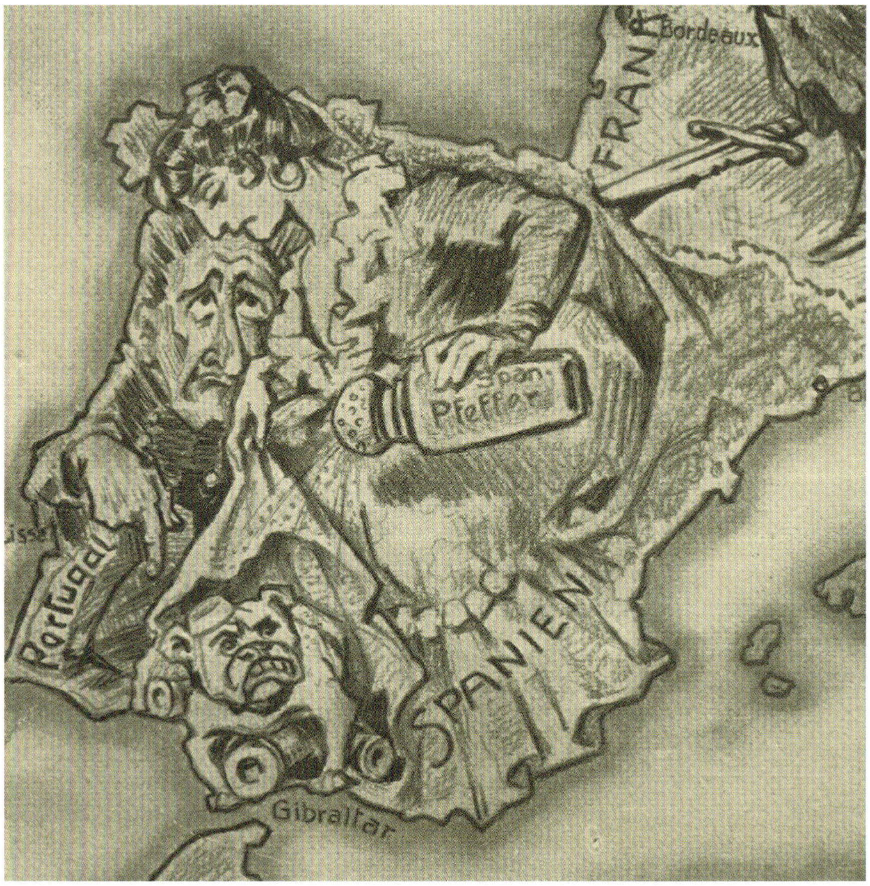

The main offensives planned and executed through the second half of 1915 and first half of 1916 were instead on the Russo-Turkish front in the Caucasus. These were triggered by Serbia's collapse, Bulgaria's entry into the war on the side of the Triple Alliance in October 1915, the routing of the Serbian army and the Entente's retreat from the debacle of Gallipoli in January 1916.[80]

On the Eastern Front, Paul von Hindenburg's shadow falls across the path of the Russian steamroller which is in full retreat. In the Balkans we see the mixed forces of Austro-Hungary, Germany and Bulgaria sweeping a wild-eyed Serb out of Albania.

Just behind a German staff car being welcomed into Turkey, we see the Berlin-Istanbul railroad. This was the first stage of an ambitious plan to drive a rail link from Berlin all the way to the port of Basra on the Persian Gulf, intended to give Germany access to the Indian Ocean and beyond.[81]

The German painter and illustrator Richard Eddelbüttel was born in Harburg in 1856. He studied at the Hamburg School of Crafts (Hamburg Gewerbeschule) before moving to Weimar in 1881, where he attended the Weimar Academy of Fine Arts (Weimar Kunstschule) and shared a studio with the expressionist painter Christian Rohlfs. In 1884 Eddelbüttel went on to study at the Berlin Academy of Arts (Berlin Akademie der Künste), and exhibited there in the 1888 Berliner Akademieausstellung. He also exhibited several times in the Berliner Großen Kunstausstellung from 1899 onwards and in the Münchener Kunstausstellung from 1911 onwards. He died in Harbug in 1942.

[80] Jukes, Geoffrey, 2002, *Essential Histories: The First World War, The Eastern Front 1914-1918*, Oxford, p. 38.

[81] Engdahl, William, 1992, *A Century of War: Anglo-American Oil Politics and the New World Order*, San Diego, CA., p. 34-35.

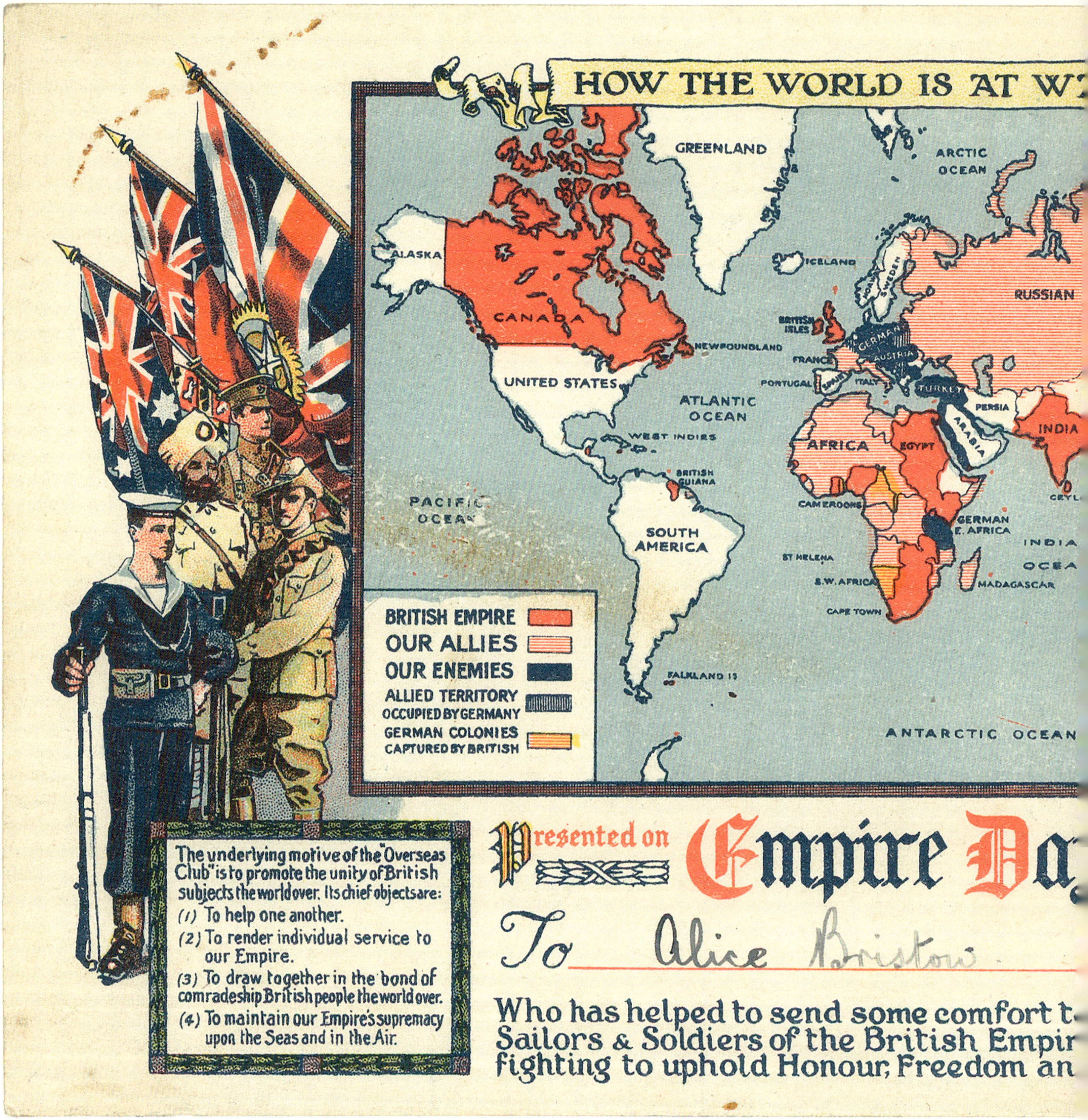

Overseas Club: How the World is at War, Presented on Empire Day

Overseas Club: How the World is at War, Presented on Empire Day, 1916

lithographed map, Sir Joseph Causton & Sons, Ltd., London, 1916, 8 x 5 in. (20 x 13 cm)

This interesting little presentation cartograph was part of the British propaganda effort to engage their entire Empire in the struggle with Germany, both in Europe and around the world.

The central map stresses the disproportionate size of the territory controlled by the Allied Powers over that of 'Our Enemies'. No doubt because of the lack of anything positive to report from the bloody stalemate on the European battlefronts, the Commonwealth successes in Africa are given unusual emphasis.[82] The German territories of Namibia, captured in 1915, and the Cameroons, captured in 1916, are shown and even, if one looks carefully, the tiny German colony of Togoland bloodlessly overrun at the beginning of the war. The only African colony still holding out is German East Africa.[83]

[82] There is a subtle semantic difference between the phrases 'Allied Territory Occupied by Germany' with its suggestion of transience and the more emphatic 'German Colonies Captured by British'.

[83] The German defence of their colony of East Africa under the command of the heroic General Paul von Lettow-Vorbeck was one of the most extraordinary campaigns of the war. Despite being entirely cut off, vastly outnumbered and reliant on captured enemy guns and supplies, Von Lettow-Vorbeck and his devoted African Askari troops managed to fight the British forces to a virtual standstill in a brilliant guerrilla war. He was the only First World War German general never to suffer a defeat, and he only accepted a ceasefire when the news reached his tiny reduced force that Germany had signed the Armistice four days earlier. See: Anderson, Ross, 2004, *The Forgotten Front, The East Africa Campaign, 1914-1918*, London.

En 1788 MIRABEAU disait déjà :
LA GUERRE est l'Industri
IRLANDE
SLESVIG HOLSTEIN
annexé en 1864
ANGLETERRE
HOLLANDE
HANOVRE
annexé
en
1866
MECKLEMBOURG
BRANDEBO
BELGIQUE
WESTPHALIE
et
PROVINCES
RHÉNANES
annexées
en
1815
BRUNSWICK
SAXE, dém
en 181
HESSE-NASSAU
annexé
en
1866
SAXE
BAVIÈRE
VERDUN
ENVAHIE
IL Y A
QUARANTE-SEPT ANS
L'ALSACE-LORRAINE
NE DIFFÈRE PAS
DES
DÉPARTEMENTS FRANÇAIS
ENVAHIS
IL Y A
TROIS ANS
DÉCLARATION DU GOUVERNEMENT FRANÇAIS
(CHAMBRE DES DÉPUTÉS 25 OCT. 1917)
ALSACE-LORRAINE
arrachée
à la France
en 1871
Gd DUCHÉ DE BADE
WURTEMBERG
SUISSE
AUTR
ESPAGNE
ITALIE
A la veille de la Guerre
la puissante Association Pangermaniste
ALLDEUTSCHER VERBAND
déclarait partout :
«il faut que le PEUPLE ALLEMA
«s'élève comme un PEUPLE DE MAI
« au dessus des PEUPLES INFÉRI
Attaqués,
Nous ne faisons que nous défendre au nom d
P.J. GALLAIS ET Cie_ EDIT. IMP. PARIS (ATELIER MAURICE NEUMONT) VISÉ No 10.514

Neumon (Maurice): La Guerre est l'Industrie Nationale de la Prusse

Neumon (Maurice):
La Guerre est l'Industrie Nationale de la Prusse [Warfare is Prussia's National Industry]

lithographed map, P. J. Gallais et Cie, Paris, December 1917, 30 x 24 in. (76 x 60 cm)

This is another very powerful French piece of wartime propaganda evoking the expansionist intentions of the German foe, once again using the image of the grasping octopus. The head of the octopus wearing a Pickelhaube helmet is centred on Prussia, while its tentacles extend from Russia to Turkey and from Spain to Britain.

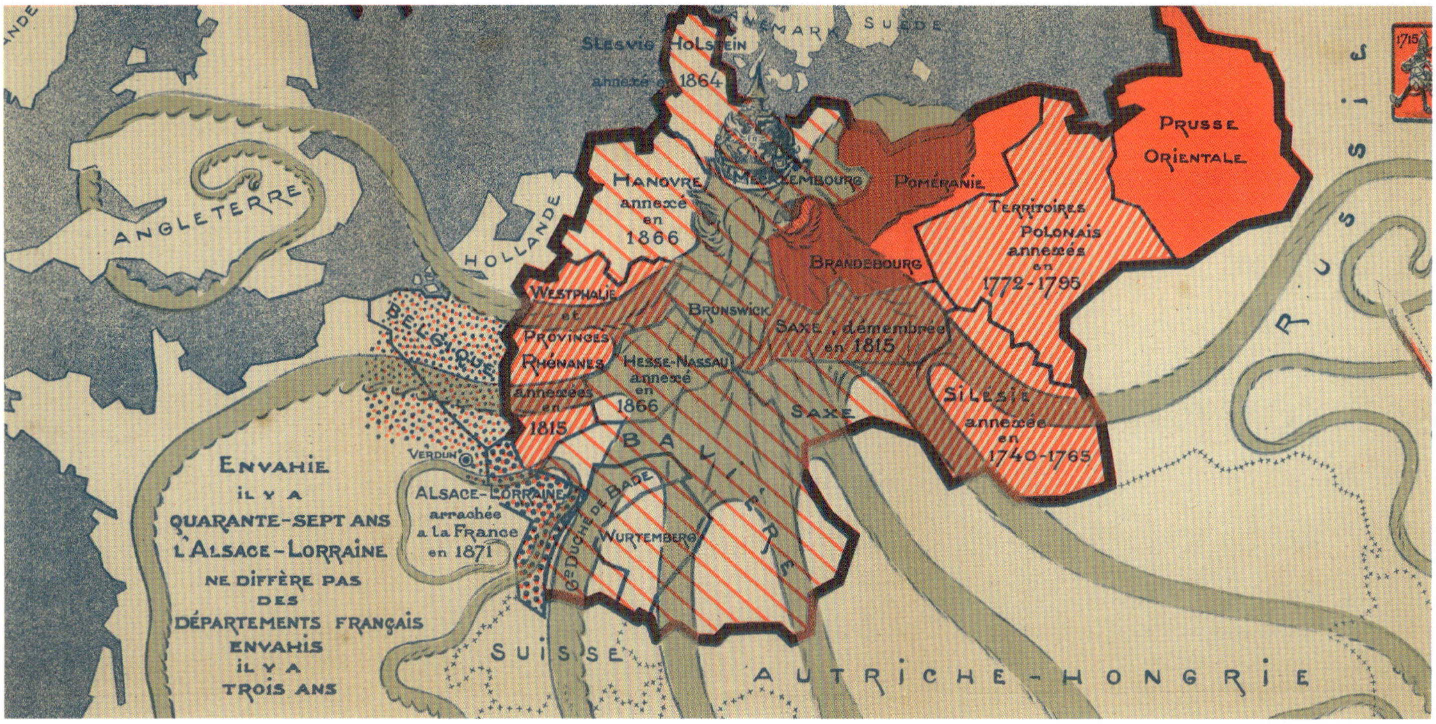

Reflecting the title "Warfare is Prussia's National Industry" a quotation from the French revolutionary Mirabeau in 1788,[84] Prussia's expansion is shown from the conquest of Silesia and Poland in the eighteenth-century, the war against Denmark over Schleswig-Holstein in 1864, and the seizing of Alsace-Lorraine from France in 1871. A dotted area in Belgium and Northern France represents the area currently occupied by the German Army.

On the right, figures denotes the growing size of the Prussian army from a small comical caricature in 1715, to the menacing giant of 1914. A chilling detail on the latter is the bloody hand print on the soldier's tunic.

The quote at the bottom, "Attacked, we can do nothing but defend ourselves in the name of Liberty and ensure our own survival" is by General Philippe Pétain. A French hero of the First World War, he was known as the 'Lion of Verdun' before gaining lasting infamy as the leader of the collaborationist Vichy Government in the Second World War.

There is also a disturbing quote by the Alldeutscher Verband (All-German League), an extreme right-wing, ultra-nationalist, racist and anti-Semitic organisation who stated shortly before the war that "The German people must rise like the Master Race, over the inferior peoples of Europe". [85]

The artist Maurice Neumon produced several famous posters during the course of the First World War, his most celebrated being *On Ne Passe Pas! 1914-1918* [None Shall Pass], depicting a resolute French soldier on the battlefield. He worked as a painter and commercial poster artist until his death in 1930. He was awarded the French Légion d'honneur for his propaganda work.[86]

[84] Honoré Gabriel Riqueti, Comte de Mirabeau, was a revered hero of the French Revolution. On his death in 1791, the Pantheon in Paris was established as a mausoleum for great Frenchmen with Mirabeau as its first occupant. Unfortunately for Mirabeau, his secret dealings with the royal family and the payments he received from Austria came to light the following year, and in 1794 his body was unceremoniously ejected, thrown into an anonymous grave and replaced by that of the murdered Jean-Paul Marat. Doyle, William, 2002, *The Oxford History of the French Revolution*, Oxford, p. 283.

[85] The Alldeutscher Verband espoused many of the most extreme policies later adopted by the National Socialists.

[86] Osterwalder, Marcus, 1983, *Dictionnaire des illustrateurs (1800-1914) – Illustrateurs, caricaturistes et affichistes*, Paris, 1983.

Was von der Ente

wenn sie Ernst machte mit dem „Selbstbestimmungs

Klimesch (F.): Was von der Entente übrig bliebe...

te übrig bliebe,

echt" ihrer eigenen Völker und die Zügel losliesse!

Verlag, Lithg. und Druck von Dietrich Reimer (Ernst Vohsen)
Berlin SW. 48, Wilhelmstr. 29

Klimesch (F.):
Was von der Entente übrig bliebe...
[What would remain of the Entente...]

coloured lithographed map, Dietrich Reimer (Ernst Vohsen), Berlin, circa 1918,
38 x 27in. (98 x 68cm)

This striking German propaganda map stresses the hypocrisy of American and other European Powers attempts to stifle Germany's own colonial ambitions.

The title in full reads: "What would remain of the Entente if it took seriously the 'rights of self-determination' of its own nations and let go of the leash!" Focus is on Germany's perception of the imbalance of colonial territories, and the double standard of the other Powers calling for greater national self-determination while continuing to keep firm control over their own overseas territories.

German resentment over their lack of colonies compared to their European rivals, Britain and France, was a major factor behind the international tensions that led to the First World War. Kaiser Wilhelm II's desire to create an overseas empire and the formation of the German High Seas Fleet in 1907 as a means of achieving this put Germany on a collision course with the British Empire.

The creation of the Triple Entente between Britain, France and Russia, also in 1907, mentioned in the title of the map, was regarded by both sides as a check on German ambition and ratcheted up hostilities yet further. [87]

[87] Tucker, Spencer (ed.), 2005, *The Encyclopaedia of World War I: A Political, Social, and Military History*, Santa Barbara, CA.; Olson, James Stuart & Shadle, Robert (eds.), 1991, *Historical Dictionary of European Imperialism*, Westport, CT., p. 20. The often confusing nomenclature of the opposing sides in the First World War altered depending on the ever-changing alliances throughout the war. At the beginning of 1914 the principal factions were the Entente powers of Britain, Russia and France, and the Triple Alliance of Germany, Austro-Hungary and Italy. The Triple Alliance was dissolved at the beginning of the war when Italy left and joined the Entente powers forming what became known as the 'Principal Allied Powers' consisting of Britain, France, Russia, Italy and also Japan. Secondary members of the Triple Entente, also collectively called the 'Allied Powers' included Belgium, Greece, Portugal, Romania, Serbia. When America declared war on Germany in 1917, it became an 'associated power' rather than a formal ally of France and Britain in order to avoid foreign entanglements. After the departure of Italy, Germany and Austria formed the 'Central Powers' ('Mittelmächte'). When Ottoman Turkey joined the alliance in 1915 and Bulgaria in 1916, this became the 'Quadruple Alliance'.

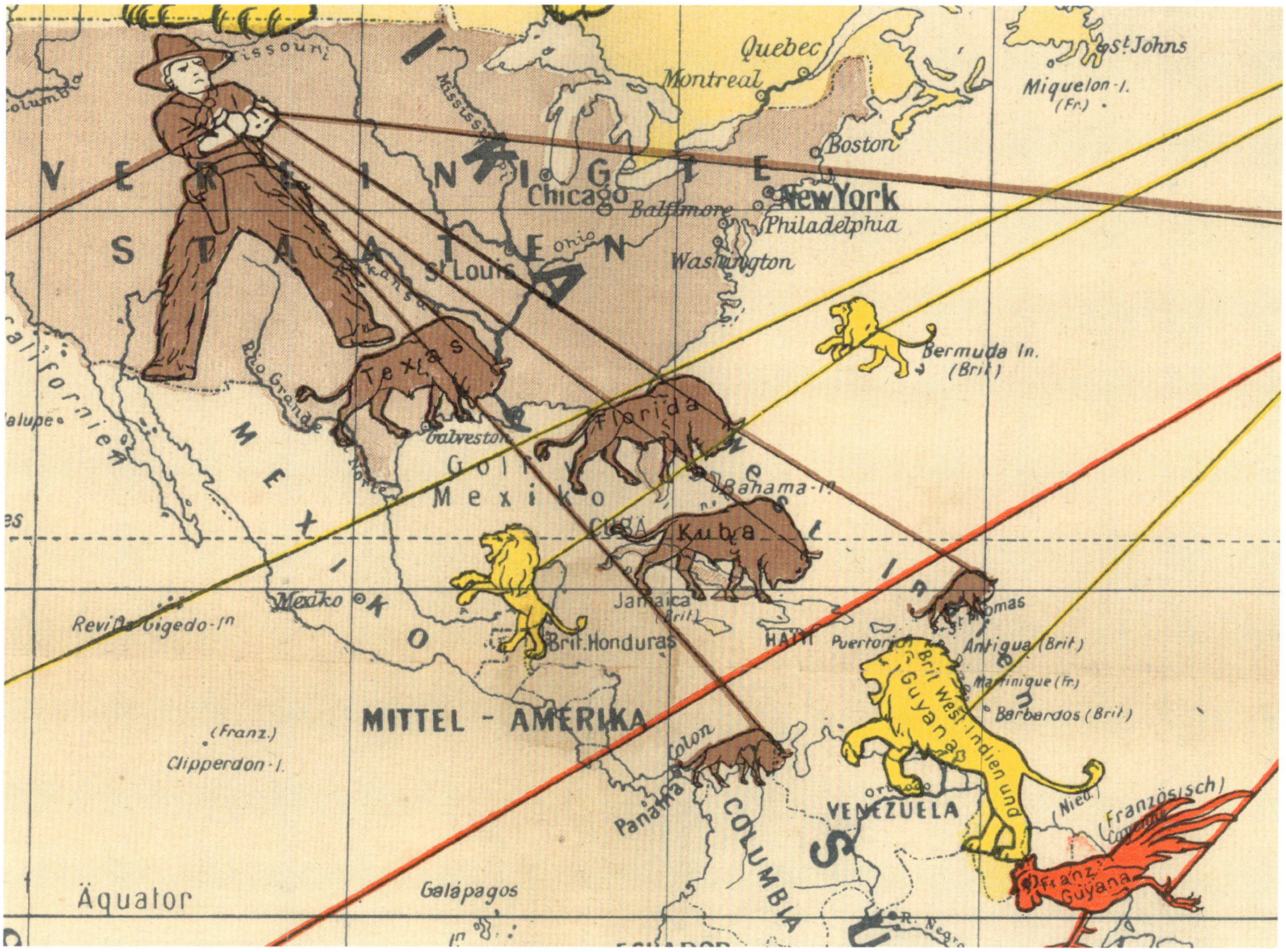

While the colonial assets of Russia, Britain, France and the U.S. are prominently shown on this map, Germany's own possessions in East Africa and their territorial concessions in China are more modestly indicated with a discreet *(D)*. Slightly more conspicuous are Germany's most recent Pacific possessions, including Samoa acquired in 1900.

No doubt in order to emphasise the inequality of territorial possessions, rather unexpectedly both Texas and Florida are shown as colonies held on leashes by America. By 1918 even the majority of Texans considered themselves an intrinsic part of the United States.[88]

[88] See: Kahn, David, 1967, *The Codebreakers*, New York. In January 1917, British cryptographers deciphered a telegram sent by Arthur Zimmerman, the German Foreign Secretary offering Mexico, in the event of a German victory, possession of New Mexico and Texas if they declared war on the U.S. The widespread publicity this received in North America after the British gave the text of the telegram to President Wilson helped push the U.S. into declaring war on Germany on 5 April 1917.

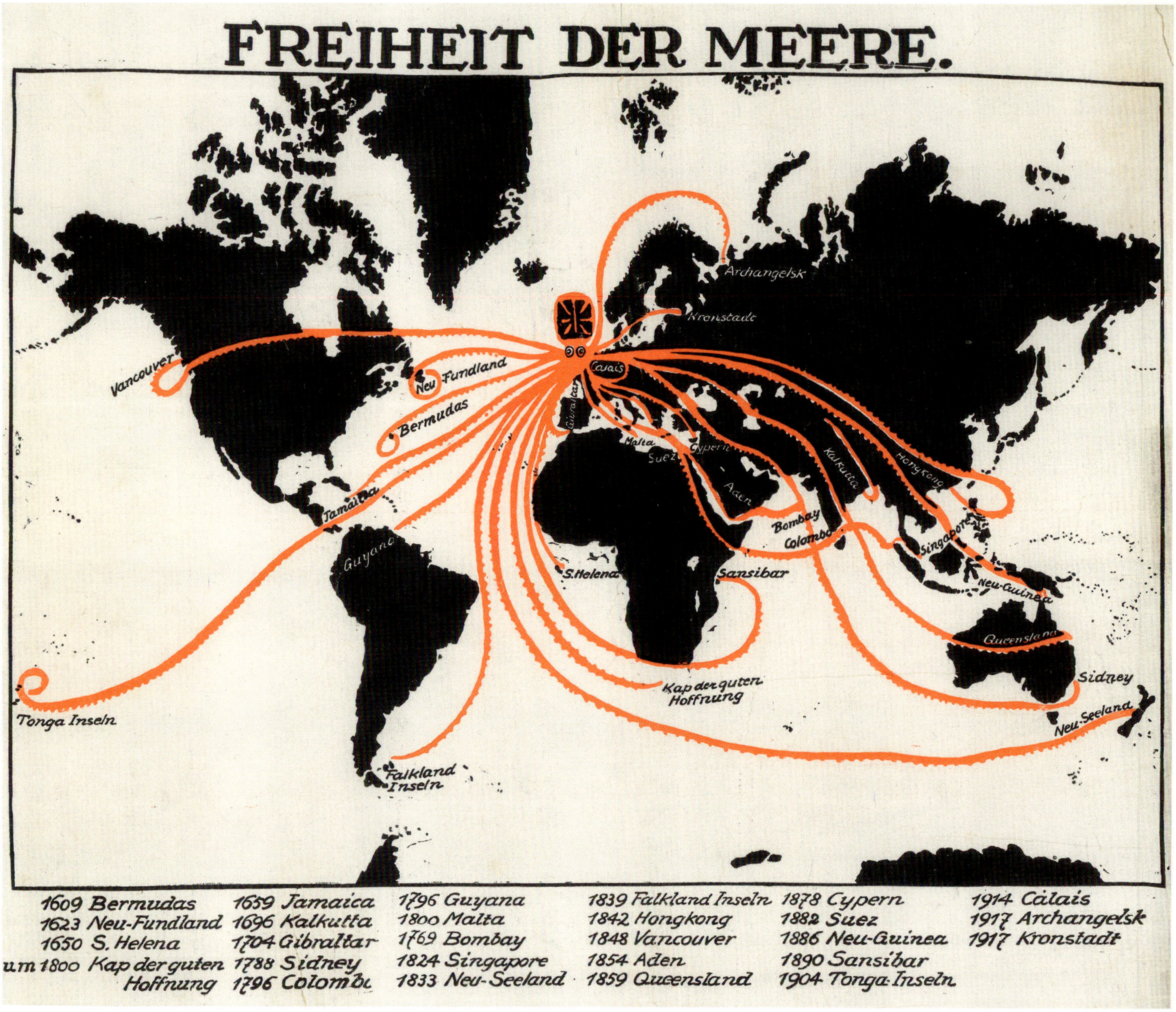

Militärische Stelle des auswärtigen Amtes [MAA]: Freiheit der Meere, [Freedom of the Seas]

lithographed map, Dietrich Reimer, Berlin, 1918, 19½ x 16 in. (49½ x 41 cm)

This stylized German caricature map again uses a grasping octopus, this time to symbolize Britain's malevolent imperialistic ambitions. An exaggerated number of tentacles stretch out to seize strategic ports around the world. Below the map is a list of significant areas conquered or colonised by Britain.

[89] Vogel, Walter, 1941, Die Organisation der amtlichen Presse- und Propagandapolitik des Deutschen Reiches von den Anfängen unter Bismarck bis zum Beginn des Jahres 1933, *Sonderheft der Monatschrift für internationale Zeitungsforschung, Zeitungswissenschaft (16), 8/9*, Berlin, pp. 29-30; Schmidt, Klaus Werner, 1982, Überwachungs-Reglementierungen im Ersten Weltkrieg (1914-1918), in Fischer, Heinz-Dietrich (ed.), *Deutsche Kommunikationskontrolle des 15. bis 20. Jahrhunderts*, Munich, pp. 185-205, p. 189.

[90] Eversdijk, Nicole, 2010, *Kultur als politische Werbemittel*, Münster, p. 53.

On 8 January 1918, President Woodrow Wilson gave a speech outlining 'The Fourteen Points', a blueprint for a post-war world. The second of these Fourteen Points was to give the 'Freedom of the Seas' to all nations in both war and peace, a clause to which both Britain and France refused to subscribe as indeed, ironically in the context of this map, did Germany.

The list of places controlled by the British includes Calais (with the date 1914) presumably because it was one of the main disembarkation points for their army in France. Also noted are Archangel and Kronstadt, both dated 1917, referring to the naval operations of the British North Russian Squadron who were attempting to support and supply the wavering military forces of first Tsar Nicholas and then his successor, the Russian Provisional Government and to encourage them to continue the fight against the Germans.

The Militärische Stelle des Auswärtigen Amtes [MAA] (Military Office of the Foreign Ministry) was created on 1 July 1916 under the leadership of Colonel von Haeften.[89] At first it handled relations with representatives of the press from neutral and enemy states, but from spring 1917 it took on a broader remit controlling all information originating from the front, occupied areas and neutral foreign states. Acting as an official censor working through its subordinate *Kunst- und Passabteilung* (Arts and Pass Department) it issued permits allowing artists to circulate their work. In October 1918 another department under the mantle of the MAA was created, the Länderabteilung zur Beobachtung der Eigenen und Fremden Propaganda (National Department for Observation of our Own and Others Propaganda). This controlled the dissemination of all printed material such pamphlets, books and maps.[90]

Briggs (W. F.):
A Merry Xmas - Allied Offensive 1918, 17th Division, August 23rd to October 11th, 1918

lithographed map, 9 x 6 in. (23 x 15 cm), W. F. Briggs, 9th Duke of Wellington's Regiment, 17th Division, 1918

This cheerful Christmas pamphlet with its depiction of a British tommy striding across the Western Front was produced to maintain morale in the closing days of the First World War. It represents the Allied breakthrough at the Second Battle of the Somme, the final decisive battle of the Western Front.

In the spring of 1918, General Erich Ludendorff, commander of the German forces, was determined to launch a major offensive using the forces released from the Eastern Front by the collapse of Russia into civil war, and before the American Expeditionary Force arrived to reinforce the war-weary Allied Powers.

The offensive initially caught the British by surprise and the Germans penetrated both the first and second lines of trenches before, the difficulties of resupply forced them to halt. They had advanced nearly forty miles, captured over 70,000 prisoners and had made one of the single largest territorial gains since the early months of the war. However the the Allied Line was bent but not broken and the exhausted and over-extended Germans were vulnerable to counterattack.[91] By the end of March, the Allies had assembled sufficient troops and a combined force of British, Canadians and Australians launched a counter offensive and rolled the Germans back to their starting positions on the Hindenburg Line. The Allied advance of forty miles, all the way to the defensive line of the Canal du Nord, is represented by the giant stride of the exultant soldier. The momentum of both advance and retreat gathered pace, and by the end of October the Allies had breached the defences of the hitherto impregnable Hindenburg Line, and forced the German High Command to concede that the war was lost.[92]

[91] Hosch, William L., 2009, *World War I: People, Politics, and Power*, New York, p. 137.

[92] Stedman, Michael, 2001, *Advance to Victory 1918: Somme*, Barnsley, Chapter 2.

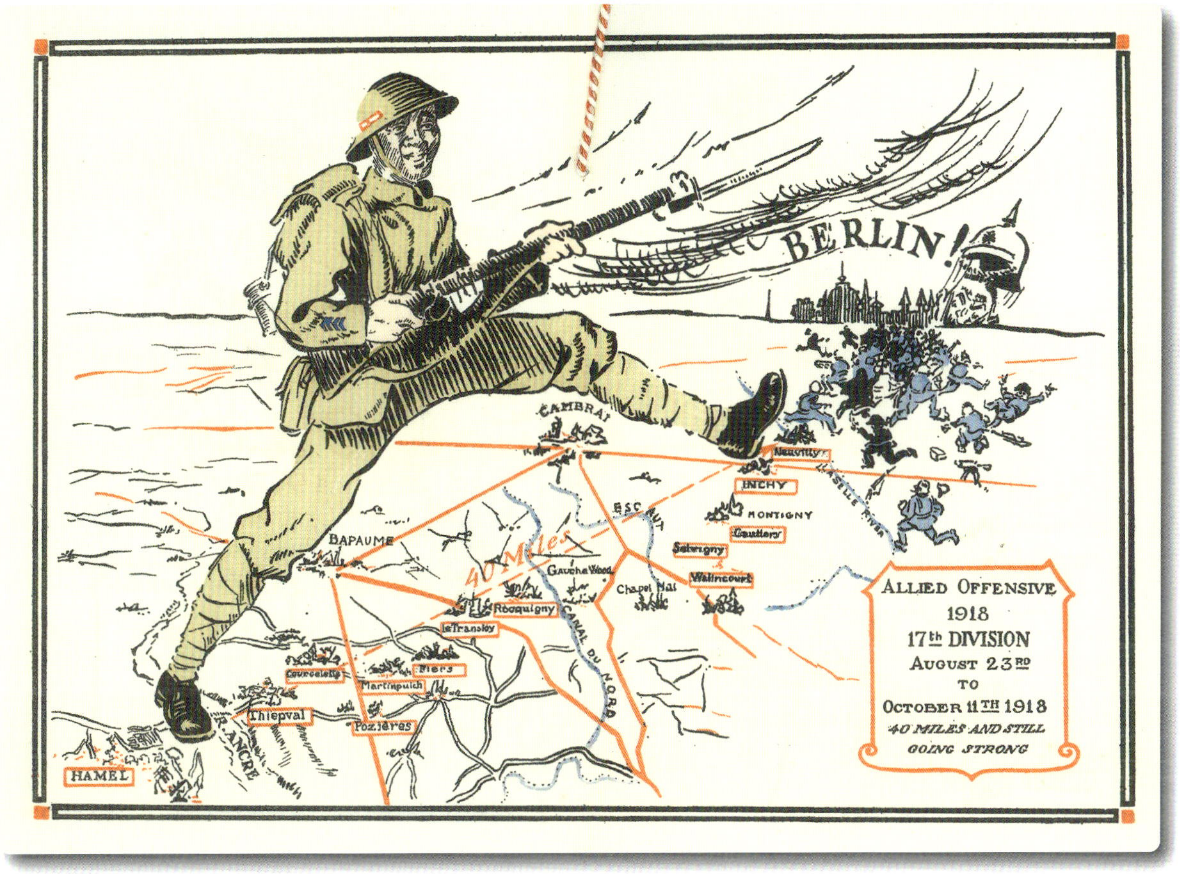

Briggs (W. F.): A Merry Xmas - Allied Offensive 1918, 17th Division, August 23rd to October 11th, 1918

The map celebrates this final triumphant phase of the campaign and shows the Germans fleeing pell mell towards Berlin and a startled Kaiser. The message in the title "40 Miles and Still Going Strong" indicates that it was drawn at the end of the offensive but before the declaration of the Armistice on 11 November 1918. The final destination shown beneath the heel of the Tommy's boot is Neuvilly on the Selle River. The fighting in what became known as the Battle of the River Selle continued on beyond 11 October (the date in the title) and indeed there was still fighting in the area in early November.[93]

The artist, Private W. F. Briggs, served in the Duke of Wellington's Regiment, part of the 17th Division. The Division landed in Boulogne, France, on 15 July 1915 and played a key role in the trench warfare in Flanders and would see heavy action in the First Battle of the Somme in 1916, at Arras and Ypres in 1917 and eventually during the Second Somme in 1918.[94]

The cover of the leaflet has an evocative sketch of a soldier lighting a cigarette, his tired expression illuminated by his match. It is perhaps not too fanciful to see on his face the glimmer of dawning hope – this time it really was going to be over by Christmas!

[93] Grehan, John & Mace, Martin (eds.), 2014, *Western Front 1917-1918*, Barnsley, pp. 184-190.

[94] Hilliard Atteridge, A. 1929, *A History of the 17th (Northern) Division*, London.

A MERRY XMAS
W.F.B.

СОВДЕПІЯ.
АРХАН ГЕЛЬСКЪ
ЯРОСЛАВЛЬ
МОСКВА
РЯЗАНЬ
ПЕТРОГРАДЪ
ПСКОВЪ
СМОЛЕНСКЪ
КУРСКЪ
ЛУГАНСКЪ
РИГА
ЧЕРНИГОВЪ
ВИЛЬНА
ГРОДНО
КІЕВЪ
ПОЛТАВА
ЕКАТЕРИНОСЛАВЪ
ЮЗОВО
НИКОЛАЕВЪ
ХЕРСОНЪ
ОДЕССА
ВАРШАВА
ЕКАТЕРИНОДАРЪ
НОВОРОССІЙСКЪ
СЕВАСТОПОЛЬ

C. A.: [Russian Civil War]

C. A.: [Russian Civil War]

chromo-lithographed map, 1919, 23 x 16 in. (58 x 40 cm)

An extraordinary and rare pictorial survival of a map that depicts the Russian Civil War from the White Russian perspective. Presumably printed in early 1919, it shows the Bolsheviks beleaguered on all sides by an international coalition of their foes.[95]

After the October Revolution of 1917, Russia descended into chaos with the Bolsheviks, monarchists, nationalists, separatists, socialists and anarchists among others desperately vying for domination. By the spring of 1919, the principal territory controlled by the Bolsheviks had shrunk to the area between Petrograd and Moscow with a corridor running down to the Caspian Sea.[96]

To the west the Poles and the Finns, the latter with German help, are attempting to gain national independence from their Russian overlords. In the north General Nikolai Nikolayevich Yudenich's army is threatening Petrograd, while British and American forces hold Arkhangelsk.[97] The Bolsheviks are depicted as ragged panic-struck peasants, one of whom is incongruously clutching a pair of presumably looted spats.

The southern forces are denoted by figures representing the British (a strikingly cheerful kilt-wearing Scotsman) and two Cossacks.[98] Material support from the Allied Powers is indicated by a British sailor passing over a field gun and a tank.[99]

It is from the east and Siberia that the most diverse of the anti-Bolshevik combatants are advancing. White Russian forces under Supreme Ruler and Commander-in-Chief, Admiral Alexander Kolchak,[100] are being supported by figures representing the Americans, the British, the Japanese and the Czechoslovak Legion.[101]

In October 1919 the Northwestern Army under Yudenich attempted to capture lightly defended Petrograd. Yudenich's potential allies, the Finns under General Gustaf Mannerheim, refused to back the assault because of Admiral Kolchak's refusal to recognize Finnish independence. Despite this, the attack started well and the White Russian forces reached the city's suburbs but failed to take the all-important Moscow-Petrograd railway. This allowed the Bolsheviks to pour reinforcements into the city which was now under the personal command of Leon Trotsky.[102] By the end of October the assault had faltered, and by early November the White Russians were in full retreat.

Belying the prominence in this map of foreign military aid for the White Russians, this support was only ever half-hearted and it dwindled to virtually nothing after the end of the First World War. The largest foreign contingent was the Japanese, with over 70,000 men, but they were more interested in territorial gain rather than helping elements of their traditional enemy.[103] The disunity and mutual distrust between the disparate forces grouped under the White Russian flag, each with their own and often conflicting war aims, would lead to their inevitable piecemeal destruction by the ruthless and single-minded Bolsheviks.

[95] The text surrounding the map identifies all the major protagonists involved in the Russian Civil War and identifies the figures on the map as follows: Finns – General Yudenich's Army – Poles; above the map they read: Lenin's and Trotsky's Government – the Northern Army and their Allies (English, Americans, etc.) – Admiral Kolchak's Army/(Soldiers from) Orenburg, the Urals – Englishmen, Americans, Japanese; on the right: Siberia/Admiral Kolchak's Government; and below the map the translations are: Englishmen in the Caspian Sea – Volunteer Army – Englishmen in the Caucasus – Volunteer Army/Don, Kuban and Terek Cossacks – Help from Allies' Volunteer Army – Southern Russia Armed Forces.

[96] At the outbreak of the First World War St. Petersburg was renamed Petrograd to make it sound less Germanic. It became Leningrad in 1924.

[97] The 'Northern Intervention' by the Entente Allies is a sometimes overlooked aspect of the First World War. After the Russian armistice with the Central Powers, the British and Americans seized Arkhangelsk to offer logistical support to the White Russians, with the hope of re-opening the Eastern Front against the Germans.

[98] The British were giving special support to the brutal Terek Cossacks, fighting as part of General Denikin's White movement army in the Caucasus.

[99] In fact only six tanks were supplied by the British and these were deployed on the Arkhangelsk Front.

[100] Smele, Jonathan D. 1996, *Civil War in Siberia: The Anti-Bolshevik Government of Admiral Kolchak, 1918-1920*, Cambridge, p. 1.

[101] A large number of Czechs elected to fight for the *Entente* Powers in the hope that in the event of the defeat of the Austro-Hungarian Empire, Czechoslovakia would be rewarded with independence. Over 60,000 Czechs would ultimately fight in Russia and at one point they controlled virtually the whole Trans-Siberian Railroad.

[102] Ziemke, Earl F. 2004, *The Red Army, 1918-1941: From Vanguard of World Revolution to America's Ally*, London, pp. 18-19.

[103] It was not until 1922 that the Bolsheviks managed to eject the last of the Japanese forces from Siberia.

Stiles (Ezra C.): Historical Map of the American Expeditionary Force

Stiles (Ezra C.): Historical Map of the American Expeditionary Force

lithographed map, Herbick & Held Printing Co., Pittsburgh, 1932, 27 x 34 in. (69 x 86 cm)

Published in 1932, this pictorial map of Western Europe pays tribute to the American Expeditionary Force (AEF) involved in the First World War. While it is a historical map looking back at the First World War, its power as a propaganda piece lies in its reminder to the American people of the last time they successfully became engaged in a European conflict.

The American Eagle hovers above the title, while cloud-born infantry men and machine gunners look down on the map. The side-borders are made up of the badges of the combat divisions, while at the bottom are those of the army corps. The density of detail is remarkable, with inset maps of Paris and the Atlantic, as well as two of the most famous battles involving the expeditionary force: the St. Mihiel Salient and the Meuse-Argonne Offensive.

The map is enlivened with numerous details and vignettes such as the grave of Quentin Roosevelt,[104] Theodore Roosevelt's youngest son, and an arrow pointing at Ypres captioned the "First Gas Attack".[105]

[104] Roosevelt, a popular and flamboyantly daring pilot was shot down behind enemy lines during a dog fight on 14 June 1918. The Germans, deeply impressed that the U.S. President's son should be prepared to fight and die, acclaimed him a hero and gave him a full military funeral.

[105] The first major gas attack on the Western Front occurred during the Second Battle of Ypres on 22 April 1915. The Germans unleashed 168 tons of chlorine gas from 5,730 canisters over a six mile section of the front.

The Battle of Saint Mihiel in September 1918 was part of a plan by the American General John Joseph 'Black Jack' Pershing[106] to overwhelm the German positions at Metz and Verdun. It was the first real offensive under American leadership, but although the attack was initially successful, the problems of supply and moving artillery forward, as so often before, halted the advance and the attack on Metz was never realised. The Germans reinforced their positions and the AEF was eventually redirected towards Meuse-Argonne.

The final Allied offensive in the First World War lasted from mid-September until the Armistice on 11 November 1918, and stretched along the entire Western front. The main effort was put in at Meuse-Argonne by the AEF and the French 4th Army, and it marks the most significant victory by the Americans during the war. It was planned by Colonel George Marshall[107], who worked closely with General Pershing in AEF headquarters. As part of the Grand Offensive, they attacked at Sedan, while the British, French and Belgians attacked at other points along the Hindenburg Line. The main American drive was north of Verdun, during which they suffered over 26,000 killed during the fighting, and as such it marks the highest number of American fatal casualties ever in a single battle.[108]

The German inability to effect anything more than temporary breakthroughs anywhere along the Western Front in four years of combat, together with the American entry into the war, brought Field Marshall Paul von Hindenburg to the following conclusion:

"So I must really say that the British food blockade and the American blow in the Argonne decided the war for the allies... without the American troops and despite a food blockade... the war could have ended in a sort of stalemate." [109]

The lessons learnt by the officers and men of the AEF during the First World War were hugely important in helping the Americans build a modern army. Pershing was so impressed by the Allied tank regiments that he ordered a young cavalry officer, Colonel George S. Patton, to create the beginnings of a tank corps.[110]

Ezra C. Stiles was an American landscape architect, writer and mapmaker.[111] He was principally known for his detailed garden designs for wealthy private clients in Pittsburgh in the late 1920s.[112] Together with this map which he created with the help of historian Paul C. Bowman, he also drew two maps of Pittsburgh for the Union Trust Company, as the city was seen in 1899 and in 1939, and one of Frick Park, Pittsburgh in 1938. His eye for detail, developed for his garden plans, is apparent in his map of the Western Front, with its wealth of references, symbols and visual features.

[106] In July 1944, Charles de Gaulle paid his respects to the now senile General Pershing in Washington. When Pershing asked after his old friend and colleague, Marshall Pétain, De Gaulle, with unusual tact, replied that when he last saw him, the Marshal was well.

[107] Later Chief of Staff of the Army during the Second World War, Secretary of State when his name was given to the Marshall Plan, and eventually Secretary of Defence. He won the Nobel Peace Prize in 1953.

[108] This compares to 7,863 deaths in total on both sides at the *Battle of Gettysburg*, the bloodiest battle of the Civil War. See: Ferrell, Robert H., 2012, *America's Deadliest Battle: Meuse-Argonne, 1918*, Lawrence, KS.

[109] Seldes, George, 1929, *You Can't Print That*, New York, p. 5; Henry, Mark & Walsh, Stephen, 2003, *The U.S. Army of World War I (Men-at-Arms)*, London, p. 4.

[110] Patton had previously served under Pershing in Mexico when they unsuccessfully chased the revolutionary and bandit, Pancho Villa.

[111] See the collection at the University of Pittsburgh: http://beta.worldcat.org/archivegrid/data/181596538

[112] Hannegan, Barry, December 1997, Ezra C. Stiles: A Re-introduction, *PHLF News*, no. 148, Philadelphia, p. 15.

Appleton (LeRoy H.): The Great Invasion of 19?? A Jingo's View, based on some Potent Realities

lithographed map, Fortune Magazine, 1935, 22 x 14 in. (56 x 35 cm)

Created by LeRoy Appleton, one of *Fortune* magazine's regular mapmakers, this map vividly illustrates the U.S.'s fear of foreign invasion during the 1930s.

The identification of possible invaders is unexpectedly broad with Japan and Britain along with *"a coalition of European powers"* all being cited as potential threats. The text lists three principal directions from which an attack on the U.S. might come. The first, presumably with Britain as the aggressor, is from Canada. The second route likely to be favoured by the coalition of European powers is from the West Indies. The third described as "the most dangerous possible international combination for the U.S. would be a coalition of Great Britain and Japan" attacking simultaneously.

This anxiety regarding foreign enemies was partly fuelled by John J. McSwain, Chairman of the U.S. Committee of Military Affair from 1934-36, and his warnings about the weakness of American armed forces.[113] This was current with a national debate on the need and form of a modern defensive air force.

A note in California is an interesting harbinger of the paranoia that would lead to the internment of over 110,000 Japanese Americans during the Second World War. The note reads: "25,000 able-bodied male Japanese settled here. Californians suppose they have secret arms and drill at night in dark halls".

Besides his work for *Fortune*, LeRoy H. Appleton also published the acclaimed book, *American Indian Design and Decoration*, which originally came out in 1950 under the title, *Indian Art of the Americas.*

THIS map illustrates three elementary attacks that could be made against the U.S.

1. An army basing on Canada could sweep down from the Northeast with the middlewestern industrial region as its objective (sometimes called the Ruhr Valley of the U.S.). This is the attack discussed in an executive meeting of the House Committee on Military Affairs, which was published accidentally and brought a reprimand to Chairman McSwain from President Roosevelt. The idea is that a coalition of powers could establish hundreds of bases for amphibian planes in Canadian waters (also in Newfoundland), so scattered that the U.S. would be unable to attack them. At a preconcerted signal they would take off for their objective (the U.S. Ruhr Valley) where 100 hits on bridges, powerhouses, oil refineries, and water-supply dams would demoralize the entire central area of the nation. Then (we now leave the McSwain hearings behind) troops could (a) attack Detroit and Buffalo, and (b) march down the Champlain corridor on Albany. If Buffalo were captured a drive could penetrate the Mohawk Valley, along the New York Central. Another, striking south, could cut the Pennsylvania, the Erie, the Baltimore & Ohio, the Chesapeake & Ohio—all the important east-west railroad lines.

2. If the enemy were a coalition of European powers he would probably launch a simultaneous attack against the Chesapeake Bay region, bringing his troops up in transports from bases in the West Indies. This attack could not succeed by itself, however.

3. The most dangerous possible international combination for the U.S. would be a coalition of Great Britain and Japan. In that event, a third attack, based on Alaska, could be launched against the Northwest. Here we are weak but *not* particularly vulnerable. Indeed, an isolated attack here or anywhere else on the Pacific Coast would be of little military importance. The strategy would doubtless be to retire while the Navy bottled the enemy from behind.

The U.S. is the safest country in the world and only its industries make it vulnerable. In the above map each smokestack represents $100,000,000 added by manufacture in 1929 (chosen because most representative of war conditions) showing the dramatic concentration of industry in the East. The red soldiers on the map show concentration points for U.S. troops—*not* permanent ones, but those which would probably be created to parry the particular attacks portrayed. The black circles with white numerals show approximate areas where new air bases are needed. These and others were granted in the Wilcox Bill, passed July 29, 1935.

[113] *Flying Magazine*, June 1934, p. 404; *Army Ordnance, vol. 15-16*, American Defense Preparedness Association, p. 368.

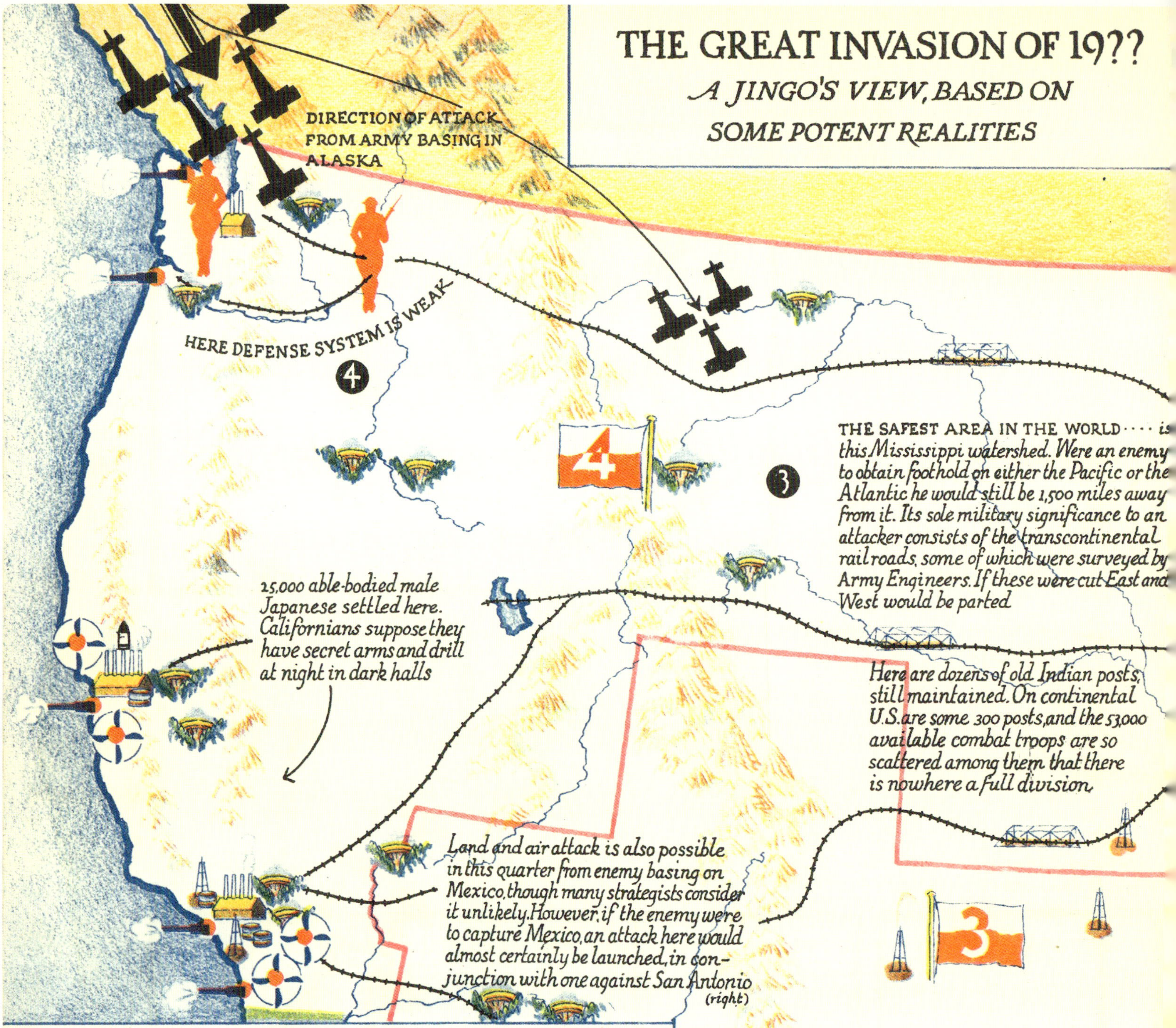

THIS map illustrates three elementary attacks that could be made against the U.S.

1. An army basing on Canada could sweep down from the Northeast with the middlewestern industrial region as its objective (sometimes called the Ruhr Valley of the U.S.). This is the attack discussed in an executive meeting of the House Committee on Military Affairs, which was published accidentally and brought a reprimand to Chairman McSwain from President Roosevelt. The idea is that a coalition of powers could establish hundreds of bases for amphibian planes in Canadian waters (also in Newfoundland), so scattered that the U.S. would be unable to attack them. At a preconcerted signal they would take off for their objective (the U.S. Ruhr Valley) where 100 hits on bridges, powerhouses, oil refineries, and water-supply dams would demoralize the entire central area of the nation. Then (we now leave the McSwain hearings behind) troops could (a) attack Detroit and Buffalo, and (b) march down the Champlain corridor on Albany. If Buffalo were captured a drive could penetrate the Mohawk Valley, along the New York Central. Another, striking south, could cut the Pennsylvania, the Erie, the Baltimore & Ohio, the Chesapeake & Ohio—all the important east-west railroad lines.

2. If the enemy were a coalition of European powers he would probably launch a simultaneous attack against the Chesapeake Bay region, bringing his troops up in transports from bases in the West Indies. This attack could not succeed by itself, however.

3. The most dangerous possible international combination for the U.S. would be a coalition of Great Britain and Japan. In that event, a third attack, based on Alaska, could be launched against the Northwest. Here we are weak but *not* particularly vulnerable. Indeed, an isolated attack here or anywhere else on the Pacific Coast would be of little military importance. The strategy would doubtless be to retire while the Navy bottled the enemy from behind.

The U.S. is the safest country in the world and only its industries make it vulnerable. In the above map each smokestack represents $100,000,000 added by manufacture in 1929 (chosen because most representative of war conditions) showing the dramatic concentration of industry in the East. The red soldiers on the map show concentration points for U.S. troops—*not* permanent ones, but those which would probably be created to parry the particular attacks portrayed. The black circles with white numerals show approximate areas where new air bases are needed. These and others were granted in the Wilcox Bill, passed July 29, 1935.

Logical direction of attack against the most vulnerable points on the U.S. map. A simultaneous attack on Detroit and Buffalo would threaten vast industrial sections
DIRECTION OF ATTACK FROM BASE IN CANADA
CHAMPLAIN-HUDSON valley is a natural military corridor joining the Mohawk at Troy
MINNEAPOLIS-ST. PAUL
DETROIT
BUFFALO
CHICAGO
ANSAS CITY
NORFOLK
FEINT
OZARK MOUNTAINS
Here the Army must still use horses
BIRMINGHAM adds only $60,000,000 a year to value of products by manufacture; hence no smoke stacks. But strategical location makes it a key to Army procurement of steel in future war
DIRECTION OF ATTACK FROM AN ENEMY BASING IN THE CARIBBEAN ISLANDS
ston
n border
FEINT ATTACKS
Army Areas
Manufacturing Arsenals
U.S. Emergency Concentration Points
Coast Defenses
Army Air Fields
Strategic Points Calling for Air Defenses
Enemy Forces
Enemy Bombers
Enemy Transports
Main Strategic Attack
Feint Attacks
Manufacturing Centers
(Each stack equals $100,000,000 added by manufacture)
Oil
Fields
Refineries
Dams
ALASKA
1500 Miles
1300 Miles
BERMUDA
2400 Miles
5300 Miles
HAWAIIAN ISLANDS
WEST INDIES
PANAMA CANAL

Editorial Office of King: [New Grand Map of Europe

Editorial Office of King:
最新欧州大地図（附列強国勢国防一覧）
キング
第一二巻第一号付録
[New Grand Map of Europe – (Includes: List of National Defences of the Great World Powers) King, vol. 12, no. 1, Supplement]

lithographed map, 42 x 29 in. (107 x 74 cm)
Dai-Nippon Yūbenkai Kodansha, 1 January 1936

This is a remarkable Japanese assessment of forces of the Western Powers on the eve of the Second World War. As with *The Daily Mail* map of 1915, the magnitude of the military and naval strength is indicated by the size of the relevant figures or objects. It is apparent from the perspective of this map that prior to the massive re-militarisation engendered by Hitler, Germany is quite the weakest of all the major powers.

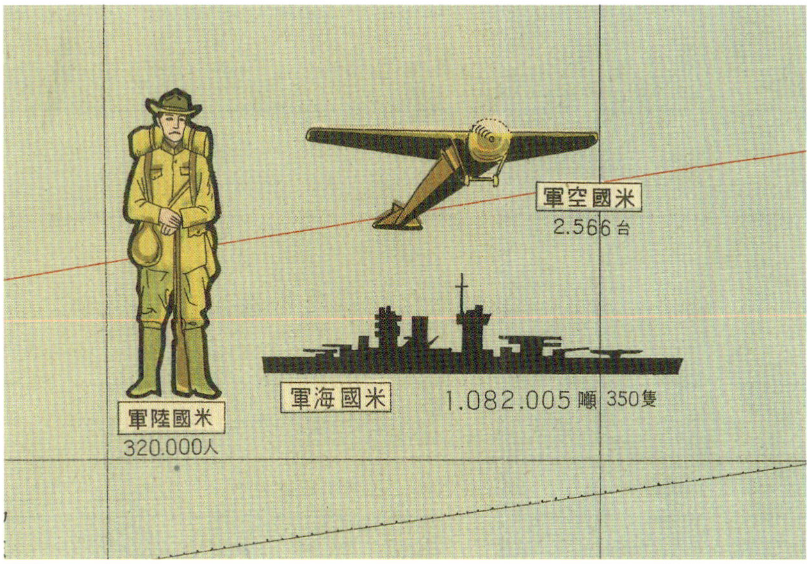

United States

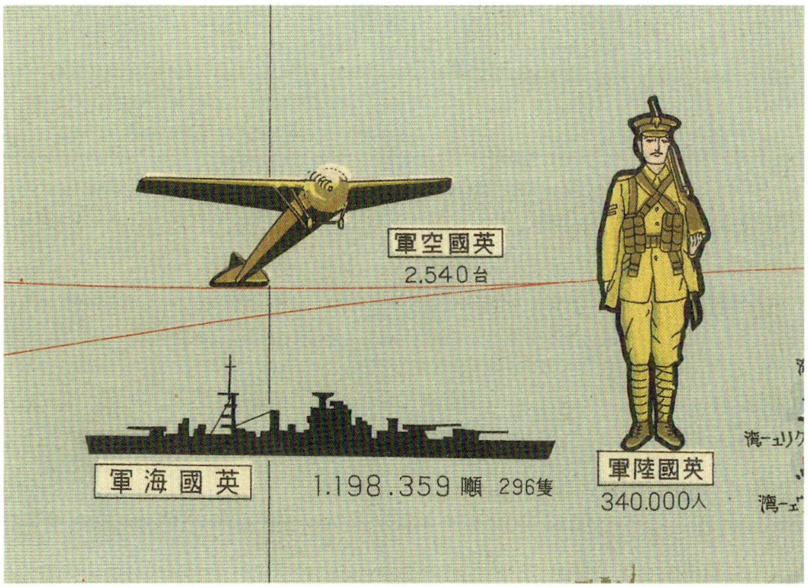

Britain

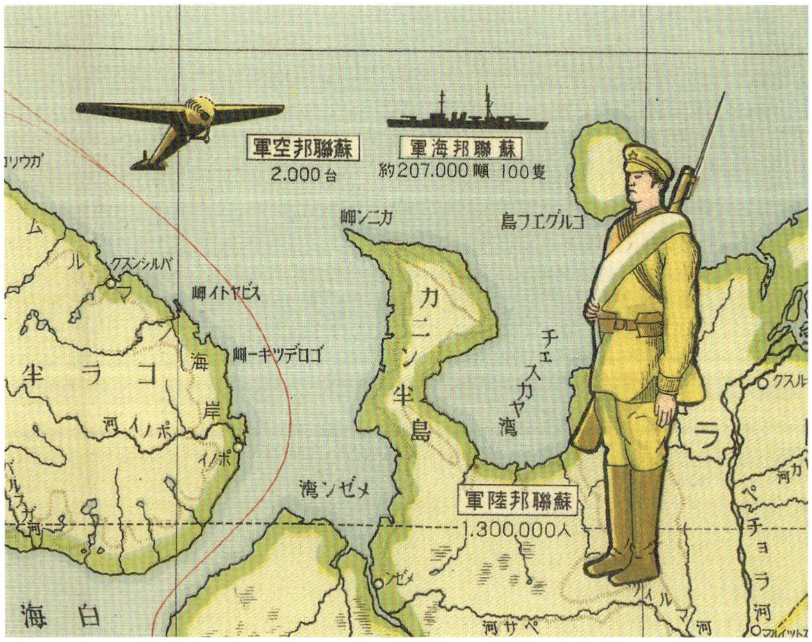

Russia

Russia's army at 1,300,000 is over twice the size of France, the next largest at 560,000. Britain has the largest navy closely followed by the U.S., while France has the largest air force with 4,675 aircraft.[114]

An interesting omission is the lack of a plane symbol for Germany as it had only officially re-established its air force in 1935. The Treaty of Versailles in 1919 forbade Germany to have either civilian or military air units. German pilots continued to train in secret in violation of the treaty, but keeping up the facade proved increasingly difficult. In 1922 with the Treaty of Rapallo, Soviet Russia and Weimar Germany, united by their mutual status as European pariahs, formed an unlikely alliance. An airfield was established in Lipetsk near the Don in 1924 for the use of German pilots flying Dutch and Russian planes under the name of the 'Fourth Squadron of the Red Air Force'.[115] This cooperation only ceased when Hitler came to power in 1933.[116] By 1939 after a massive plane building program and with combat experience gained fighting the Republicans in Spain during the Civil War, the Luftwaffe was one of the strongest air forces in the world.[117]

[114] France had on paper what appeared to be the largest and best air force of any of the combatants at the start of the Second World War but the statistics were deceptive. Only a quarter of French aircraft were operational during the Battle of France because of the poor siting of squadrons and lack of pilots. Furthermore strategic and tactical blunders and the lack of armour-piercing rounds meant that despite numerous acts of individual heroism the French Air Force was completely ineffectual in halting the German armoured advance. See: Fabyanic, Lt. Col. Thomas A., 1985, The French Air Force of 1940: Was It Defeated by the Luftwaffe or by Politics, *Air University Review September-October*, Bryn Mawr College, Pennsylvania.

[115] Budiansky, Stephen, 2004, *Air Power: The Men, Machines, and Ideas that Revolutionized War, from Kitty Hawk to Gulf War II.* Penguin Group, 2004.

[116] Hooton, E. R., 2007, *Luftwaffe at War: Gathering Storm 1933-39, vol. 1*, London, p. 30.

[117] Killen, John, 2003, *The Luftwaffe: A History*, Barnsley, p. 93.

Throughout the 1930s, Japan's rapid militarisation and her imperial ambitions in South East Asia were ratchetting up the tension in the region. This apprehension was also felt by the United States and the European powers with their many colonies and extensive interests in Asia. This map gives us a glimpse of the Japanese view of the world in the build-up to the war.

The publishing house of Kodansha was established when its founder, Seiji Noma, departed from Dai-Nippon Yūbenkai (Greater Japan Oratorical Society) in 1909. The name is taken from one of the company's early magazines: *Kōdan Club*, and was first used in 1911 when it formerly merged with Dai-Nippon Yūbenkai. For decades it was Japan's largest publishing house, but in recent years this position has been threatened by Shogakukan. This map was issued by King, one of Kodansha's subsidiary companies, and reflects the company motto: *"Omoshirokute tame ni naru"* ("To be interesting and beneficial").

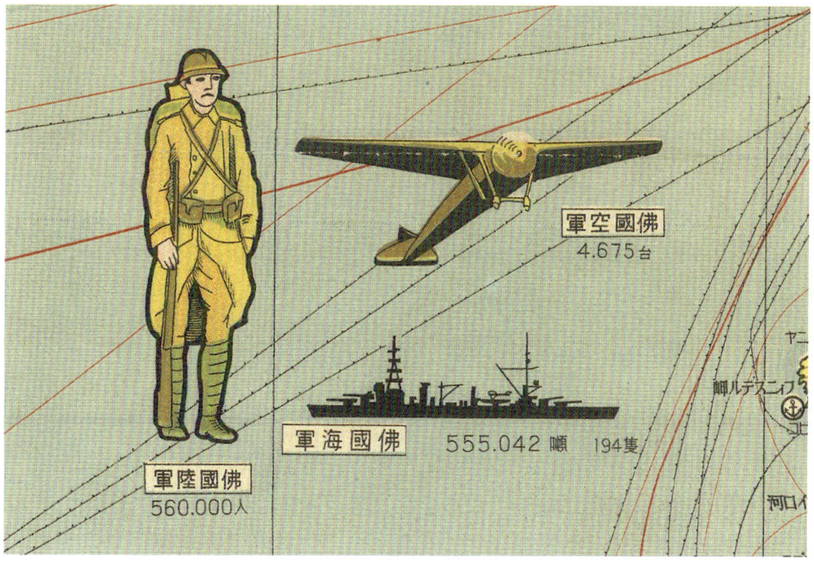

France

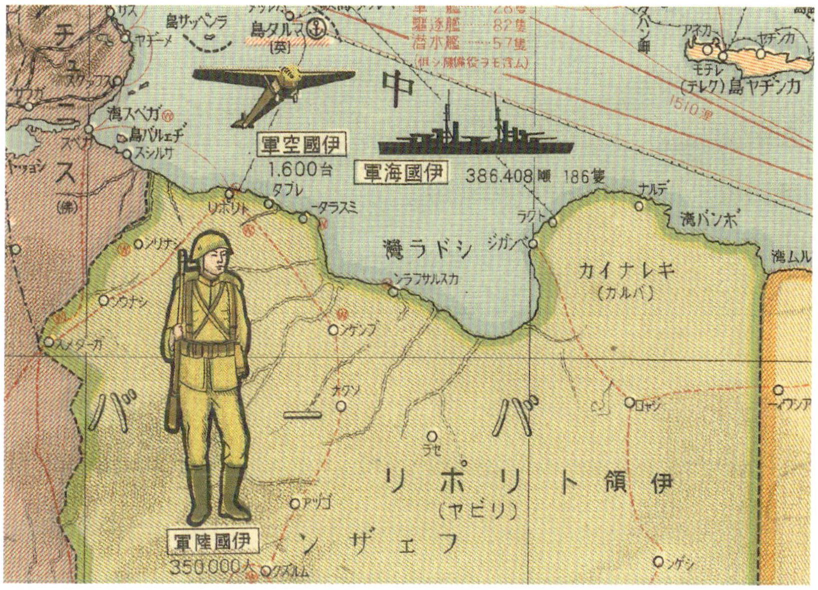

Italy

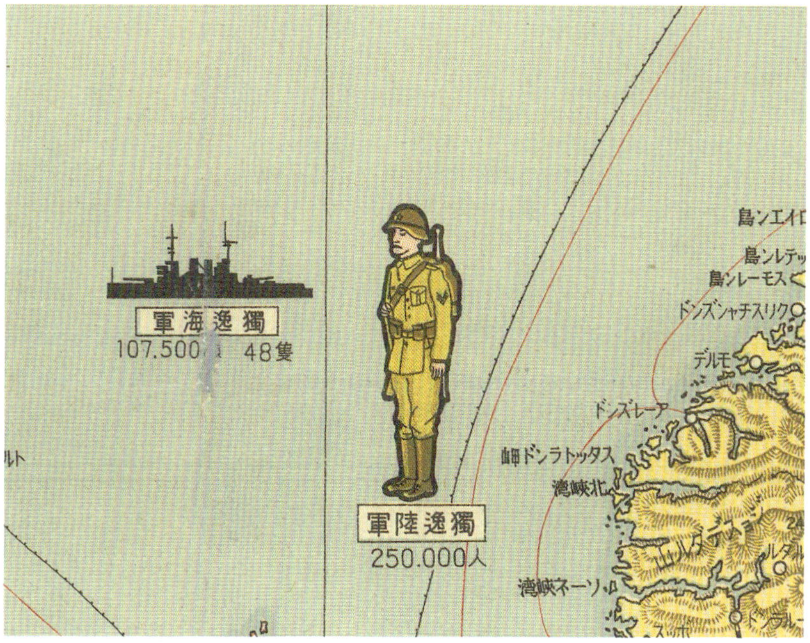

Germany

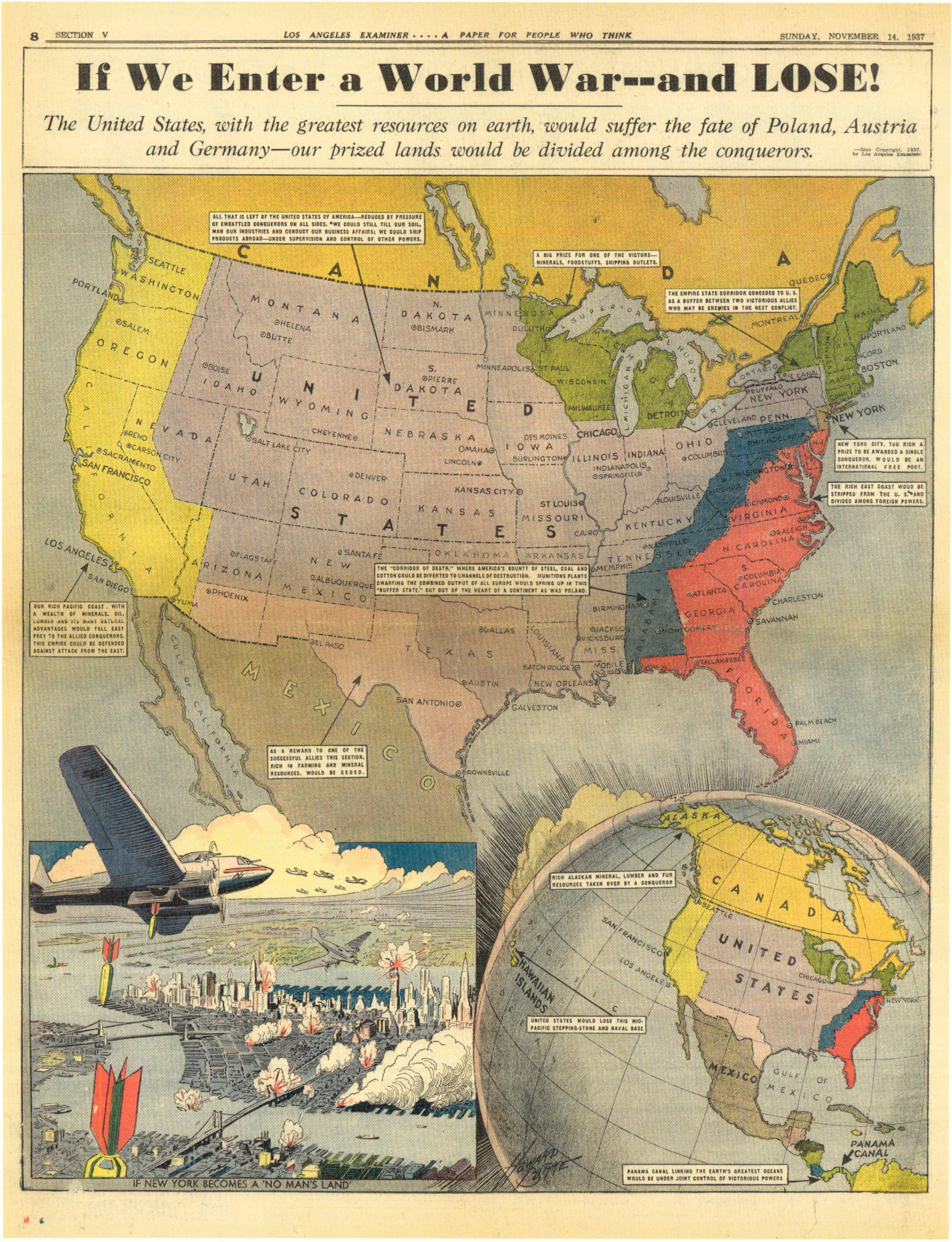

Burke (Howard): If We Enter a World War – and Lose!

Burke (Howard): If We Enter a World War – and Lose!

newspaper map, Los Angeles Examiner, Los Angeles, 1937, 17 x 22 (43 x 56 cm)

The U.S. fear of foreign invaders has taken an even darker turn in this map, which imagines the consequences of American defeat.

America is shown dismembered and apportioned among the victors leaving only a reduced U.S. in the centre of the continent. While the supposed belligerents remain unnamed and are referred to as "the victorious allies" or "foreign powers", the map's creator seems to envisage a coalition of combatants including Japan, Britain and European powers.

The large inset vignette shows New York suffering a heavy bombing raid. In the 1930s Canada is the only place in mainland North America from which bombers would be able to reach New York, and this is perhaps an indication of the paranoia underlying this doomsday scenario. The mention of a "corridor" and of Poland as a buffer state recalls the way in which Poland was referred to by nationalist Germans, as opposed to the Poles themselves and their Western allies who regarded Poland as a fulfilment of a historic dream. In turn, it suggests the political sympathies of the publishers, if not necessarily of the artist.

The *Los Angeles Examiner* was a tabloid newspaper owned by William Randolph Hearst. It embraced many of the features of 'yellow journalism' employing extravagant sensationalism and rampant nationalism often illustrated, as here, with coloured Sunday supplements. The artist, Howard Burke, was the art critic and an illustrator for the newspaper from 1932-1940, during which he produced a series of pictorial maps documenting the build-up towards war.[118]

[118] Hughes, Edan, 1986, 3rd ed., *Artists in California 1786-1940*, Sacramento, CA.

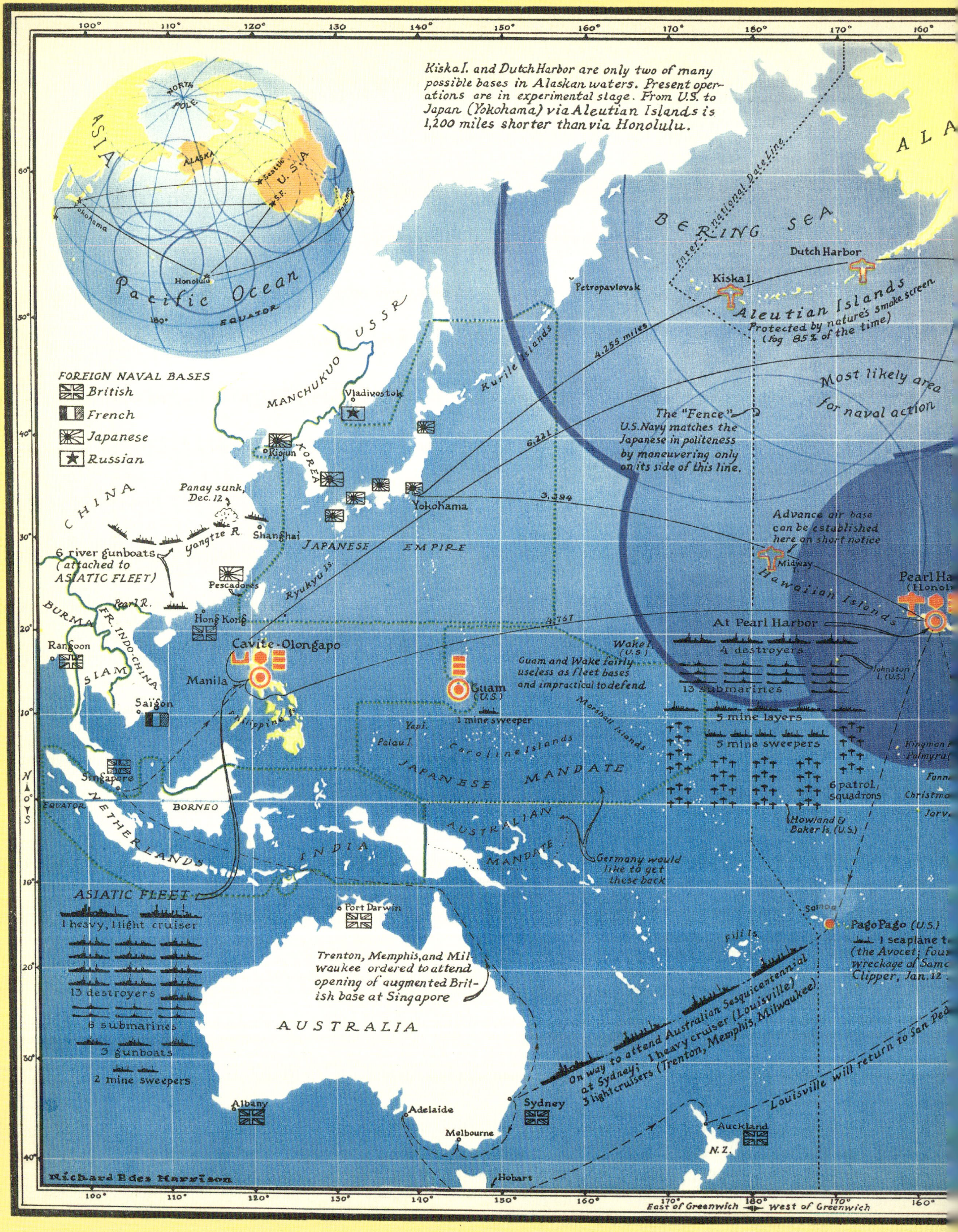

Kiska I. and Dutch Harbor are only two of many possible bases in Alaskan waters. Present operations are in experimental stage. From U.S. to Japan (Yokohama) via Aleutian Islands is 1,200 miles shorter than via Honolulu.
NORTH POLE
ASIA
ALASKA
U.S.A.
Seattle
S.F.
Panama
Yokohama
Honolulu
Pacific Ocean
EQUATOR
180°
FOREIGN NAVAL BASES
British
French
Japanese
Russian
BERING SEA
International Date Line
Dutch Harbor
Kiska I.
Petropavlovsk
Aleutian Islands
Protected by nature's smoke screen (fog 85% of the time)
Most likely area for naval action
The "Fence" U.S. Navy matches the Japanese in politeness by maneuvering only on its side of this line.
USSR
MANCHUKUO
Vladivostok
Kurile Islands
KOREA
Riojun
4,255 miles
6,221
3,394
4,767
Yokohama
JAPANESE EMPIRE
CHINA
Panay sunk, Dec. 12
Yangtze R.
Shanghai
6 river gunboats (attached to ASIATIC FLEET)
Pescadores
Ryukyu Is.
Pearl R.
BURMA
FR. INDO-CHINA
Hong Kong
Rangoon
SIAM
Cavite-Olongapo
Manila
Saigon
Philippine Is.
Advance air base can be established here on short notice
Midway I.
Hawaiian Islands
Pearl Ha
(Honolu
At Pearl Harbor
4 destroyers
13 submarines
5 mine layers
5 mine sweepers
6 patrol squadrons
Wake I. (U.S.)
Guam and Wake fairly useless as fleet bases and impractical to defend
Guam (U.S.)
1 mine sweeper
Johnston I. (U.S.)
Yap I.
Palau I.
Caroline Islands
Marshall Islands
JAPANESE MANDATE
Kingman R.
Palmyra
Fanning
Christmas
Jarvis
Howland & Baker Is. (U.S.)
Singapore
EQUATOR
BORNEO
NETHERLANDS INDIA
AUSTRALIAN MANDATE
Germany would like to get these back
ASIATIC FLEET
1 heavy, 1 light cruiser
13 destroyers
6 submarines
3 gunboats
2 mine sweepers
Port Darwin
Trenton, Memphis, and Milwaukee ordered to attend opening of augmented British base at Singapore
Samoa
Pago Pago (U.S.)
1 seaplane t
(the Avocet; four
wreckage of Samo
Clipper, Jan. 12
Fiji Is.
On way to attend Australian Sesquicentennial at Sydney: 1 heavy cruiser (Louisville) 3 light cruisers (Trenton, Memphis, Milwaukee).
Louisville will return to San Ped
AUSTRALIA
Albany
Adelaide
Melbourne
Sydney
Hobart
Auckland
N.Z.
Richard Edes Harrison
East of Greenwich
West of Greenwich

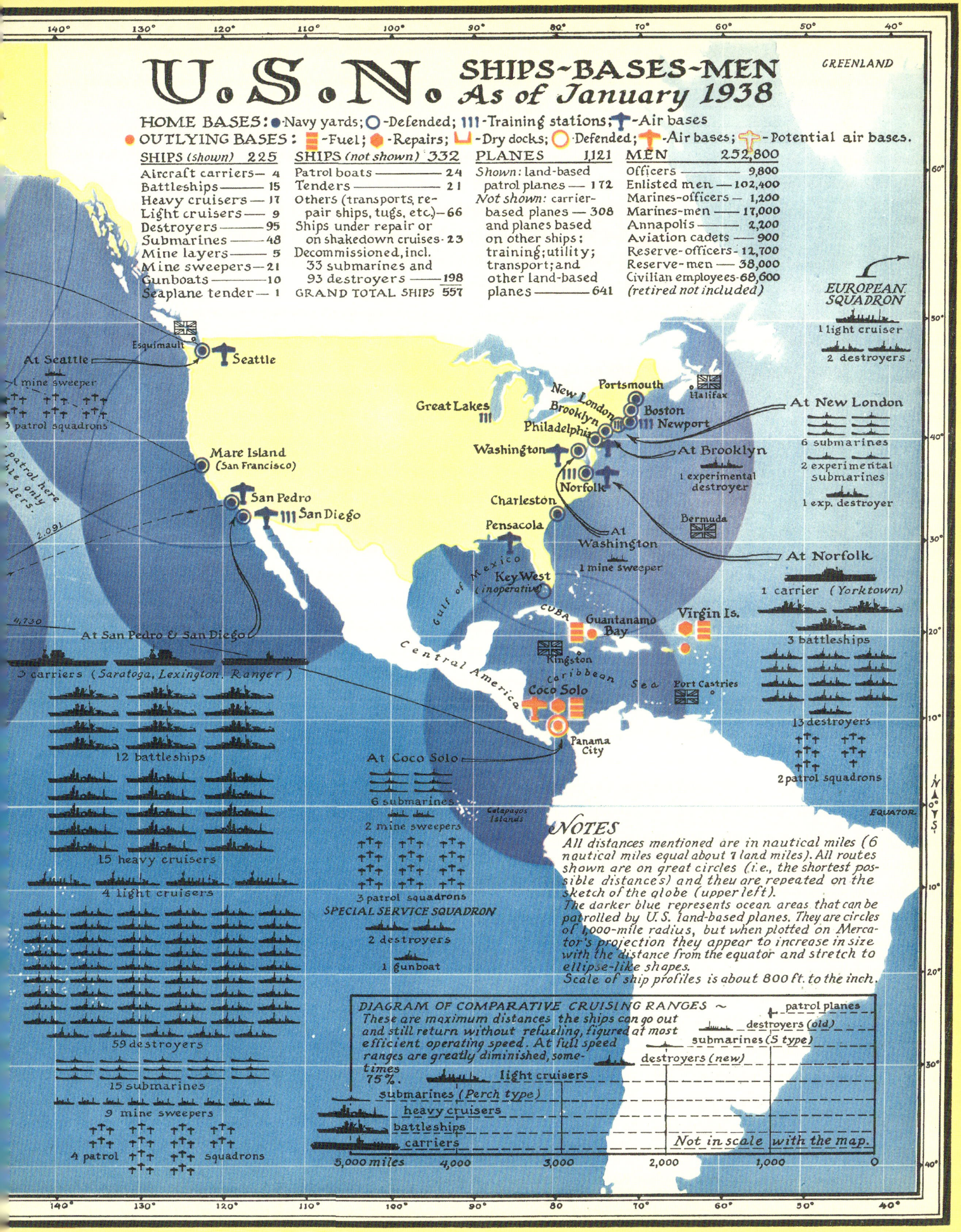

Harrison (Richard Edes): U.S. Navy - Ships, Bases, Men As of January 1938

Harrison (Richard Edes): U.S. Navy - Ships, Bases, Men As of January 1938

lithographed map, Fortune Magazine, 1938, 22 x 14 in. (56 x 36 cm)

This elegant pictorial map of the Pacific details the bases and strength of the U.S. navy in the build-up towards war.

The escalation of international tension and the threat of war in the 1930s led to a massive resurgence of interest in maps, and above all in pictorial cartographic descriptions of the rapidly changing world situation. Newspapers and magazines in the U.S. competed with each other to produce eye-catching and informative maps. Some, such as many of those printed by the Hearst newspapers, had an overtly propagandist purpose while others, such as those published by *Fortune* magazine, attempted to offer their readers a more apparently factual overview of the current situation.

In 1938 the bulk of the American Pacific fleet was based in San Diego. When war with Japan began to look inevitable, it was moved to its forward base of Pearl Harbour in Hawaii. The fleet's three aircraft carriers are given as *Saratoga, Lexington* and *Ranger*. *Ranger* was the U.S.'s first purpose-built carrier and by the outbreak of hostilities she was regarded as too small for the Pacific Theatre and was transferred to the Atlantic and replaced with the *Enterprise*.[119]

[119] The *Saratoga*, *Enterprise* and *Ranger* were the only three carriers built before 1938 to survive the war.

The "Fence", the arbitrary line that divided the U.S. and Japanese navy's areas of operation, is shown, as is an area described as "Most likely area for naval action". Presciently the latter is placed near the site of the Battle of Midway, the decisive battle of the naval war in the Pacific.

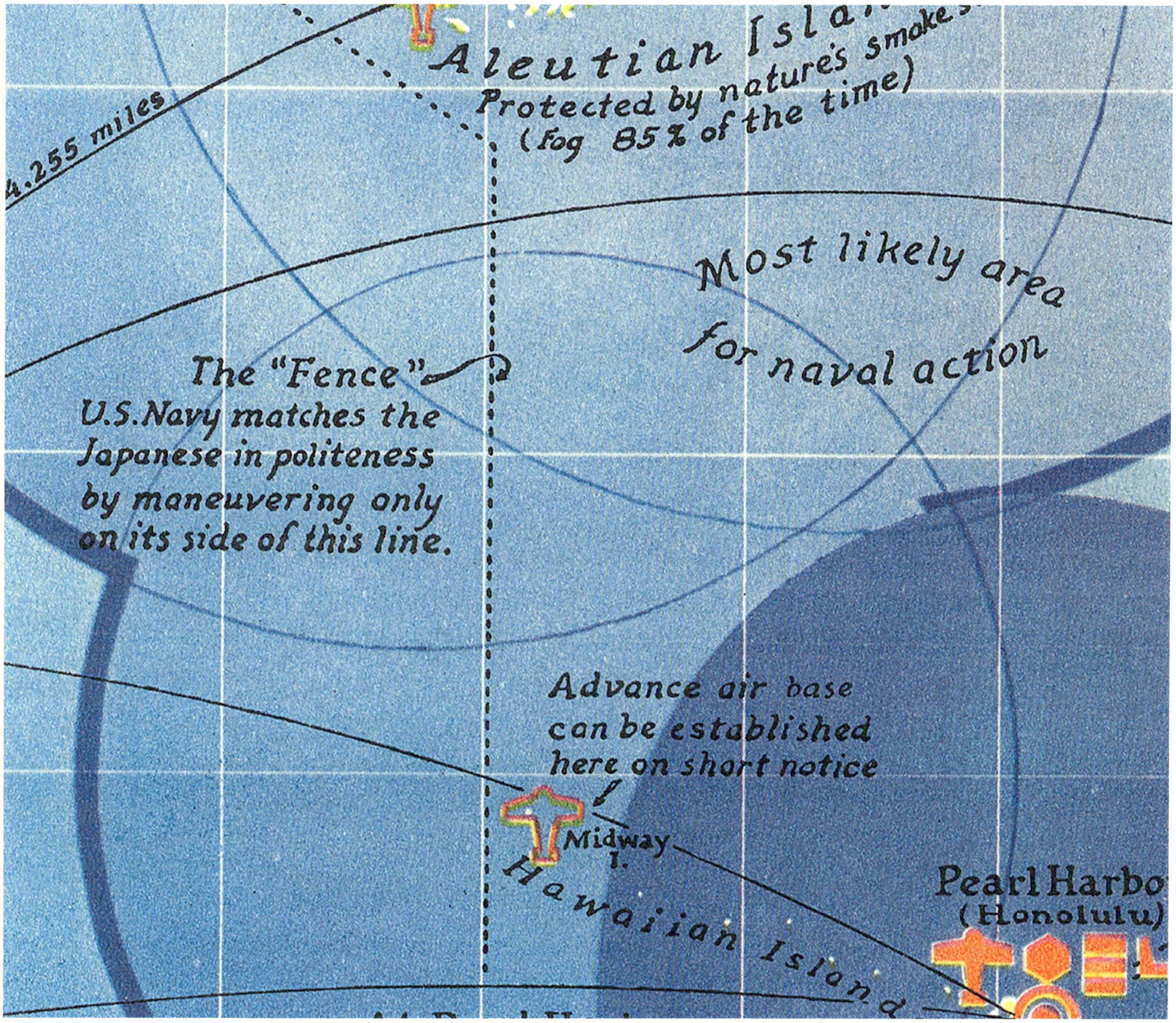

Richard Edes Harrison is one of the most important pictorial mapmakers to illuminate the conflicts of the Second World War. His greatest innovation was to realise that the inevitable distortions involved in maps using Mercator's Projection often misrepresented the all-important geographical relationships between the combatant nations. Although he uses Mercator's Projection here as the most suitable for the Pacific, he notes the distorting effect it has on apparent distances and he would go on to create and use a number different projections depending on the subject [see map on pp.128-9].

The United Press and the Columbia Broadcasting System: 1939 A.D. Pictorial History and World Map

The United Press and the Columbia Broadcasting System: 1939 A.D. Pictorial History and World Map

lithographed map, Lip & Ba, New York, 1939, 30 x 20 in. (76 x 51 cm)

This is a detailed summation of the principal events leading up to the declaration of war in Europe in 1939, as seen from the neutral American perspective.

Many Americans viewed the outbreak of war in Europe with deep ambivalence, not least because of the sizable numbers of U.S. citizens of German and Italian descent. There was also a feeling that the conflict was the business of the 'Old World' not the 'New' and that Germany was not a threat to the U.S. – Japanese ambitions in the Pacific were seen as a much more immediate danger.

The central map reflects the situation following the German-Soviet Non-Aggression Pact. Poland is split between the two powers with Czechoslovakia now incorporated within the Greater Reich. Japan is shown controlling Korea, and the Japanese puppet state of Manchukuo in Manchuria is marked. Also indicated are the British and French colonies, coloured in orange-red and blue respectively, and also the major shipping lanes.

"The news events depicted were selected for their outstanding interest and importance…" and include the ending of the Spanish Civil War (26 January 1939), Nazi-Germany's invasion of Czechoslovakia (15 March), the signing of the Molotov-Ribbentrop Non-Aggression Pact (21 August), England's and France's declaration of war against Germany (3 September), Soviet Russia's invasion of Finland (30 November) and the scuttling of the pocket battleship *Admiral Graf Spee* trapped in the harbour of Montevideo, Uruguay (17 December).

Two further incidents are shown which would have far-reaching consequences and help propel the U.S. into the war. The first is the seizure of the neutral American merchant ship, *The City of Flint* as it traded with Britain. The imprisonment of its crew raised considerable anti-Nazi feelings amongst Americans. The second was the ending of the American-Japanese Trade Agreement, the so-called 'China Open Door Treaty'.[120] Designed to curb Japanese Imperial ambition, it served to further convince the Japanese High Command that war with the U.S. was inevitable.

This map, with its fascinating images of current events, shows a world already starting its inexorable slide into the abyss of war.

[120] "*The Japanese conception of the 'open door' is a door for the Japanese to come in and for foreigners to go out.*" (*The Argus*, Melbourne, 14 January 1941).

Anon: Nazi War Aims – Grab! Grab!! Grab!!!

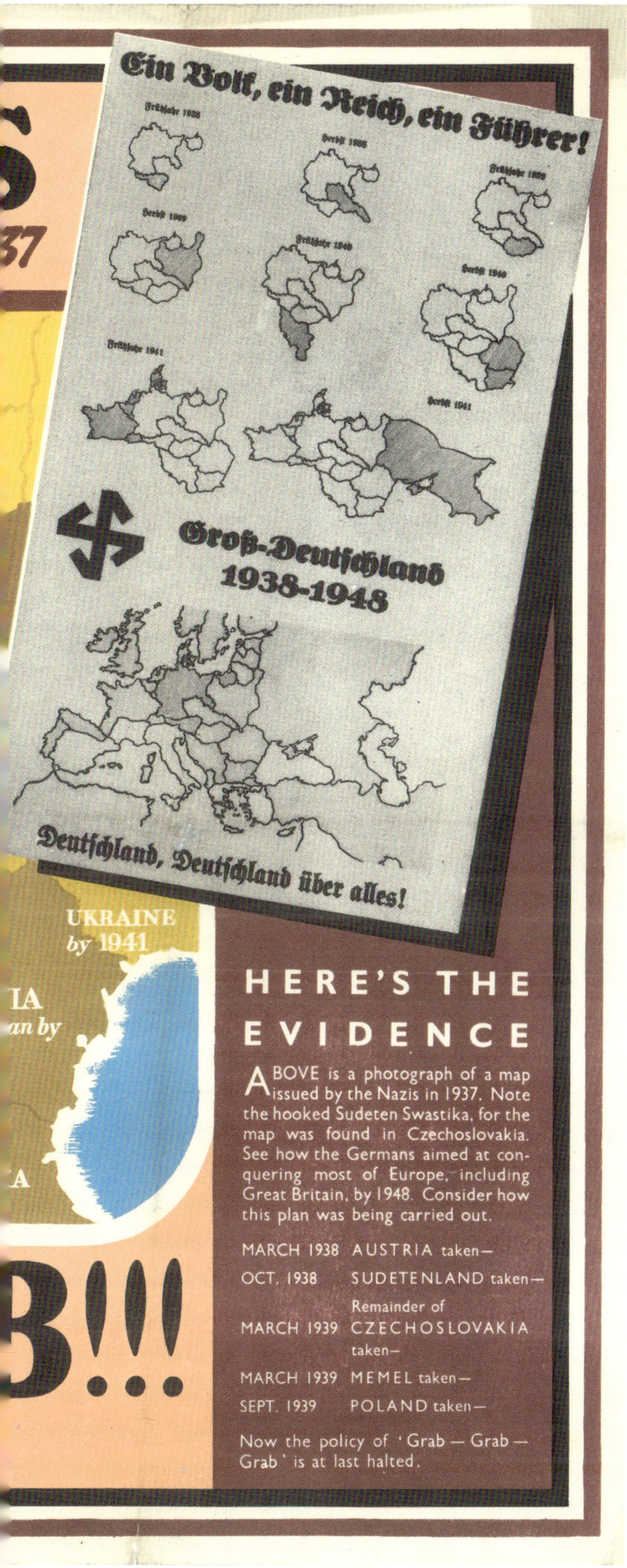

Anon: Nazi War Aims – Grab! Grab!! Grab!!!

coloured lithographed map, c.1939, 15 x 20 in. (38 x 51 cm)

This striking propaganda map offers insight into how the Allies perceived the scale of Germany's territorial ambitions.

It must have been drawn after the fall of Poland in September 1939 but before the invasion of France in May 1940. The smaller image in the top right-hand corner purports to be a facsimile of a poster detailing Nazi Germany's secret plans and their timetable for territorial expansion. While the German Army High Command (Oberkommando des Heeres, or OKH) undoubtedly had contingency invasion plans for all the marked countries, it is highly unlikely that the Nazis would ever have allowed something as incendiary as this map to be printed or disseminated. It is more probable that it was invented by the anonymous propagandist to lend credibility to his message.

While warning of the imminent threat posed by a resurgent and aggressive Germany, it is noticeable that the map's creator has, like so many, underestimated the speed of the German attacks. Northern France, Belgium and Holland had all been conquered a year before the Nazis' claimed target dates. Britain is marked for subjugation in 1948 when in fact the planned invasion Unternehmen Seelöwe (Operation Sea Lion), was scheduled for 1940.

Soviet Russia is not marked as a target. Germany and Russia had only just signed their non-aggression pact and it was obviously as inconceivable to this mapmaker as it was to Joseph Stalin, that Hitler could already be planning Operation Barbarossa, the invasion of Russia.

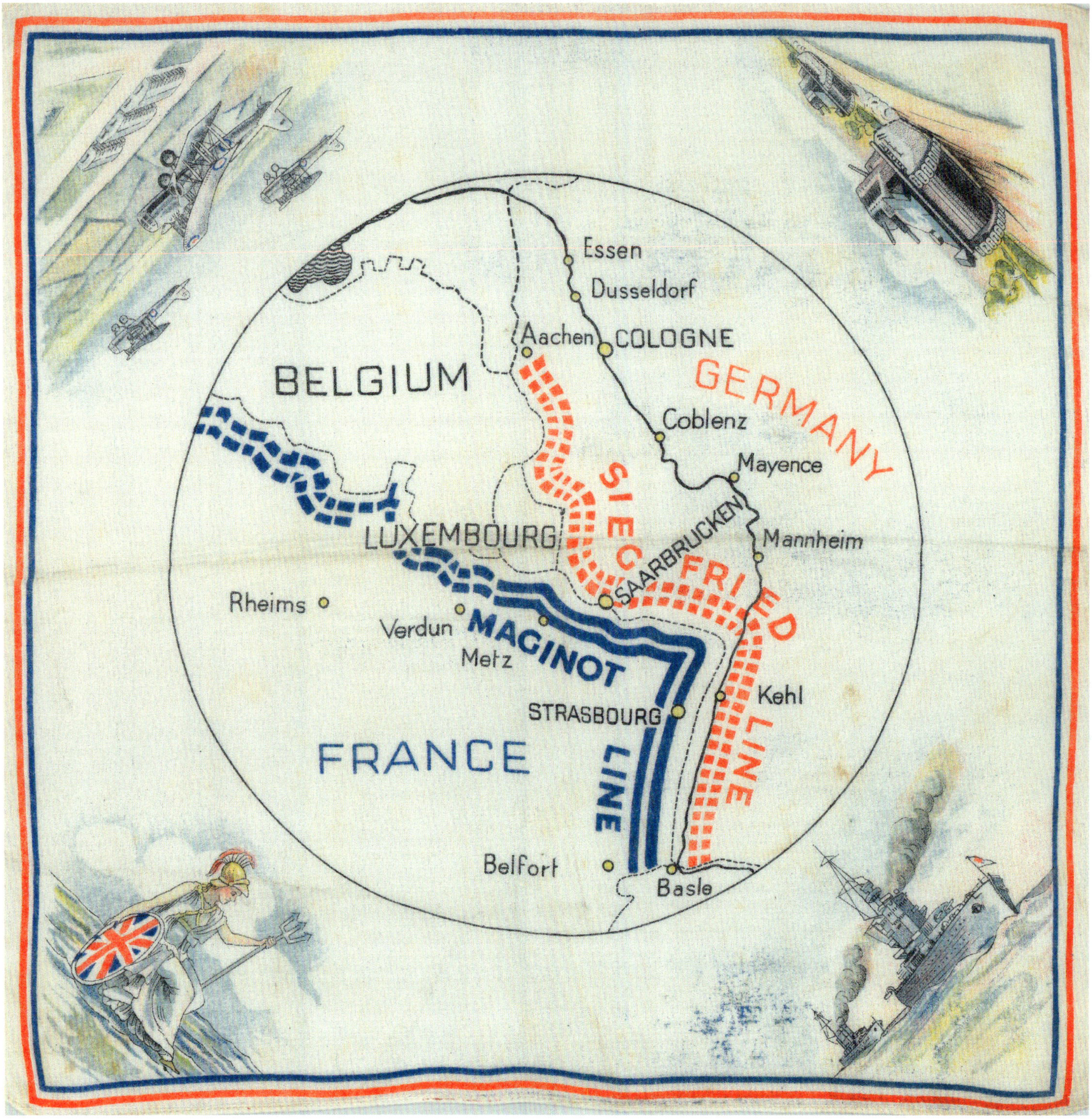

Anon: [*Siegfried/Maginot Lines*]

Anon: [Siegfried/Maginot Lines]

lithographed map printed on cloth, 1939, 12 x 12 in. (30 x 30 cm)

This rare map printed on cloth was designed to reassure the men in the British Expeditionary Force (BEF), and their families at home, of the strength of the French defences. It shows the *Maginot* and *Siegfried Lines*, two great fixed lines of fortifications between Germany and France at the start of the war. It would have been circulated during the 'Phoney War', the period between the deployment of the BEF in France in September 1939 and the German armoured breakout through the Belgian Ardennes Forest in May 1940.

It is often said that generals always go into battle prepared to fight the previous war but it could be argued that the French generals were getting ready to fight the last war but one. It was the mortifying French defeat in the Franco-Prussian War, combined with the traumatic experience of trench warfare in the First World War that convinced many of their generals of the overwhelming advantage of defence over attack.[121] The French army was considered the strongest in Europe, but as early as 1935, Marshal Tukhachevsky, Commander in Chief of the Red Army, 1925-28, and a prominent victim of Stalin's purge in 1938, predicted that a fast moving German attack would completely overwhelm an immobilised French army whose lack of radio communications rendered it unable to adapt to changing circumstances.[122]

The horrendous loss of life suffered by the French in the First World War, 1.4 million dead, 4.4 per cent of the entire population, was the highest of any of the major powers.[123] This resulted in a determination that this must never be allowed to happen again and made the concept of an impregnable defence very compelling.[124] The Maginot Line, named after the French Minister of War, André Maginot and running along the border with Germany, Luxembourg and Belgium, was just one part of a defensive system designed to enclose the whole of France.

The original plan for a hugely reinforced defensive line was first suggested by Marshall Joffre, one of the heroes of the First World War, and supported by another, Marshall Pétain, but it did not go before the Chamber of Deputies until January 1930.[125] An ambitious five year building plan was agreed with a seemingly enormous budget of 300 billion francs. Intended to create a series of linked fortresses stretching from the Mediterranean, all along France's eastern border to the North Sea, it soon ran wildly over budget with constant extensions to its planned completion date. Along the border with Italy, modernisations to already existing nineteenth-century forts were considered sufficient and the geography of Switzerland represented a natural fortress, so the main focus was put on the 471 miles from the border with Germany to the North Sea. The first part, along the Rhine with the river acting as a natural barrier, seemed secure to the planners, so it was the stretch above and west of the Rhine bordering Alsace-Lorraine that attracted the attention and the resources of the engineers. Along this historically contested land, they constructed 50 'ouvrages': vast self-sufficient concrete fortresses built underground each with its own water supply and powered by their own electricity generators. The largest housed over a thousand men and had everything to keep them self-sufficient for months including extensive wine vaults, a dentist, a jail and a morgue.[126]

The gaping weakness of the Maginot Line is apparent even from the simple plan depicted on this handkerchief. The Maginot Line proper, indicated by the solid blue line, stops at Sedan on the French Belgian border. This was primarily because of lack of funds but was justified by the belief espoused by Marshall Pétain that the great forests of the Belgian Ardennes were in themselves a natural barrier impassable to an invading force. This mistake was compounded by the French confidence that Belgium, with her army of over 600,000 and whose neutrality was explicitly guaranteed by Germany, would be able to resist any invasion. This was despite Belgium's refusal to even discuss a joint defence strategy with France as to do so might be seen to compromise her neutrality.[127]

[121] The German army drew very different lessons from the First World War. It was their experience of the success of small groups of highly mobile mixed units in circumnavigating fixed fortifications that led them to what became the theories of Blitzkrieg. See: Deighton, Len, 1979, *Blitzkrieg*, London, Chapter "*Western Front*".

[122] Tukhachevsky on the French army, *Pravda*, Moscow, 29 March 1935.

[123] Westwell, Ian & Cove, Dennis (eds.), 2002, *History of World War I, vol. 3*, New York, p. 781.

[124] Beevor, Antony, 2012, *The Second World War*, London, p. 49.

[125] Boyce, Robert W. D. (ed.), 1998, *French Foreign and Defence Policy, 1918-1940: The Decline and Fall of a Great Power*, London, p. 186.

[126] Chelminski, Rudolph, 1997, The Maginot line, *Smithsonian*, June 1997, pp. 90-100.

[127] Proctor, John, 2012, *American Resolve and the Art of War: A Study and Application of Military Tactics*, Bloomington, In., p. 104.

The cardinal error of the French planners was a failure to recognise that the increased mobility of the German mechanised divisions, combined with air superiority, rendered fixed defences redundant. German Army Group A consisting of 45 divisions including seven armoured Panzer divisions punched through the Ardennes, simply following the regular roads, bypassing the Maginot Line. Belgium, her army antiquated and out-manoeuvred, proved wholly inadequate in the face of the Wehrmacht and surrendered only eighteen days after the Germans crossed their border.[128]

The Maginot Line was a much greater catastrophe for France than just being utterly useless. The huge sums spent on it were at the expense of the regular army whose training and equipment as a consequence fell far behind that of their opponents. When the false sense of invincibility it conferred evaporated in the face of the swift German advance, the French military high command was paralysed and unable to comprehend, let alone deal with the changing circumstances. Finally 400,000 men, nearly half the permanent French army, were manning the line itself and found themselves immobilised and irrelevant as the Germans swept on past towards Paris and the coast.

The other line of fortifications marked on the map is, of course, the Siegfried Line.[129] Built in response to the Maginot Line, the Siegfried Line stretched 390 miles from Kleve, on the border with the Netherlands, to Weil am Rhein on the border with Switzerland. Unlike the French military planners, Hitler had fully grasped the changes to modern warfare and regarded the line, built between 1938 and 1940, as a Nazi propaganda tool rather than a practical fortification.[130]

The Maginot Line remains today both above and below ground, and the French government has never quite known what to do with it. It is too vast to remove or ignore and while six of the great 'ouvrages' have been turned into tourist attractions, others have been handed over to local landowners who farm around them. A few have even had houses constructed on top of them and must therefore be regarded as having the best defended and certainly the most expensive wine cellars ever built.

[128] Kaufmann, J. E. & Kaufmann, H. W., 2006, *Fortress France: The Maginot Line and the French Defenses in World War II*, Mechanicsburg, PA., pp. 145-146.

[129] In Germany it was generally referred to as *Westwall* – it was the Allies who named it the Siegfried Line after the similar line of fortifications in the First World War.

[130] Short, Neil, 2004, *Germany's West Wall – The Siegfried Line*, Oxford, pp. 4-5, 12.

Garnett (Frederick): *"Zog It Off"*

chromo-lithographed postcard, London, 1939, $3\frac{1}{2}$ x $5\frac{1}{2}$ in. (9 x 14 cm)

Garnett (Frederick): "Zog It Off".

This postcard was published shortly after the invasion of Poland and the Declaration of War but prior to the Fall of France. John Bull, equipped with a third arm thanks to Zog, a multipurpose cleaning product, is shown enthusiastically trying to wipe Germany off the map of Europe. The fact that the third arm is over Albania suggests a sly double-entendre that contemporary readers would have grasped immediately. Albania's King Zog had been driven from his throne by an Italian invasion in April 1939, months before the postcard was published.

KONIGS PLATZ
THE REICHSTAG
BRANDENBURG GATE
UNTER
Hotel Adlon
THE BRITISH EMBASSY
HITLERS NEW CHANCELLERY
HAUS
BELLEVUE ALL.
SIEGES ALLEE
HERMANN GORING STR.
TIERGARTEN
WERTHEIMS
POTSDAMER PLATZ
LEIPZIGER
WILHELM STRASSE
AIR MINISTRY
POTSDAMER STATION
SAARLAND
ANHALTER STATION
BELLE

Flying visit of Truth to
Berlin
in the form of an R.A.F.
leaflet raid here fancifully depicted
— but not forgetting a great many hard facts —
by Rex Whistler
STETTINER STATION
HOHENZOLLERN MUSEUM MONBIJOU PALACE
THE CATHEDRAL
THE PALACE
THE TOWN HALL
Lustgarten
LINDEN
KONIG STR.
GERTRAUDT STR.
THE FRENCH CHURCH
CROWN PRINCE'S PALACE
GENDARMENMARKT
STRASSE
FRIEDRICH
ALLIANCE
SS
MEIN KAMPF

Whistler (Rex): Flying Visit of Truth to Berlin

photo-lithographed map, published by Illustrated, 9th December 1939, London, 19 x 12 in. (48 x 31 cm)

This wonderful illustration mocking the Nazi leadership commemorates the first RAF raid over Berlin.

On the night of 1 October 1939, four Whitley Bombers of 10 Squadron of the RAF flew over Berlin. They could only drop leaflets as the RAF was not authorised to bomb civilian property outside the immediate warzones. This policy was rescinded after the German fire-bombing of Rotterdam in May 1940.

In the upper left corner, the winged figure of Britannia is surrounded by several chubby putti wearing flying helmets and goggles scattering bundles of leaflets.

In the lower right, Hitler, Goering and Goebbels react to the raid with anger and fear while the German Foreign Minister, Joachim von Ribbentrop cowers beneath the table. Further images of the three principals adorn the title cartouche.

Rex Whistler was a much sought-after portrait painter, muralist and a prolific and brilliant illustrator. His style, as evidenced with this piece, was characterised by quirky wit and seemingly effortless panache. Despite being 35 when the war started, Whistler was determined to join up and he was commissioned a lieutenant in the Welsh Guards. His tank was easily identifiable for the bucket he hung on the outside to contain his paintbrushes.[131] After the D-Day landings he was in Normandy with the Guards Armoured Division when he was killed by a German mortar round. He was 39.

[131] Beevor, Antony, 2009, *D-Day and the Battle for Normandy*, Toronto, p. 313.

SS
MEIN
KAMPF

Italy Hopes War Will Bring Nev

Neutral Role Nears End as Mussolini Moves to Win by Arms What F

Los Angeles Examiner

SUNDAY, JUNE 2, 1940

Screen and Drama

March of Events

WHAT ITALY WANTS

What does Italy expect to get out of this war?

Does she want Corsica, Savoy and Nice back from France? Possibly.

Does she want French Tunisia to match her Libya and with it the Anglo-Egyptian Sudan and Djibouti to add to Ethiopia and Somaliland, already hers, and to afford a base of possible later action with Germany against Egypt? Probably.

Does she want the Cameroons, French-mandated African colony, once Germany's? Perhaps. She has already said she wants a share of the Suez Canal, needed to connect her with Ethiopia.

There is reason to believe Il Duce wants Yugoslavia to fill in the gap in his sway on the Eastern Adriatic between Albania and Fiume, which would make that sea an Italian lake and give Rome at least an equal share with Berlin and Moscow in the domination of the Balkans which Italy is expected to protect from Allied invasion, especially in the Mediterranean.

Air bases in Spain and the Canary Islands are other aspirations.

Achievement of these objectives would make Italy one of the greatest world powers and restore the Mediterranean as the Mare Nostrum of the Roman Empire.

Failure to get what she wanted from the first world war made Italy a coy and enigmatic neutral in this one—a Juliet to such Romeo nations as Germany, Britain and France. Whether their wooing will be her opportunity or her undoing remains to be seen as she approaches the brink of actual hostilities.

Although she can get much needed coal from Germany and from England—at a price—and iron from Yugoslavia, Italy may have to look to red Russia for oil as well as for a suitable Balkan deal. The 43,000,000 Roman Catholic Italians don't like their government flirting with Russia.

If Italy actively joins Germany, Britain and France may try to blockade the Strait of Gibraltar and the Suez and force Turkey to blockade the Dardanelles. As Italy gets six-sevenths of her imports by sea this would be a calamity.

In case of war Italy would have to defend a long coastline against the British and French fleets and man her northwest frontier forts against French invasion through Piedmont. Recent maneuvers proved that theoretically Italy could repel a French invasion there.

Italy's navy isn't as large as France's but she has 130 destroyers and 105 submarines which she could use to harass an assailant, especially between Sicily and Libya. She has an air force of 3000 planes and 116,000 men in the active air service and 102,000 trained reserves. Her army numbers 2,240,000 men and she has a trained reserve of 5,175,000.

If she goes to war, Italy would have to defend her own shores, her African possessions and her stake in the Balkans.

If she is forced out of her neutrality, Italy will probably go with Germany, as only by combination with Germany and Russia can Italy get all that she wants.

Burke (Howard): Italy Hopes War Will Bring New Roman Empire

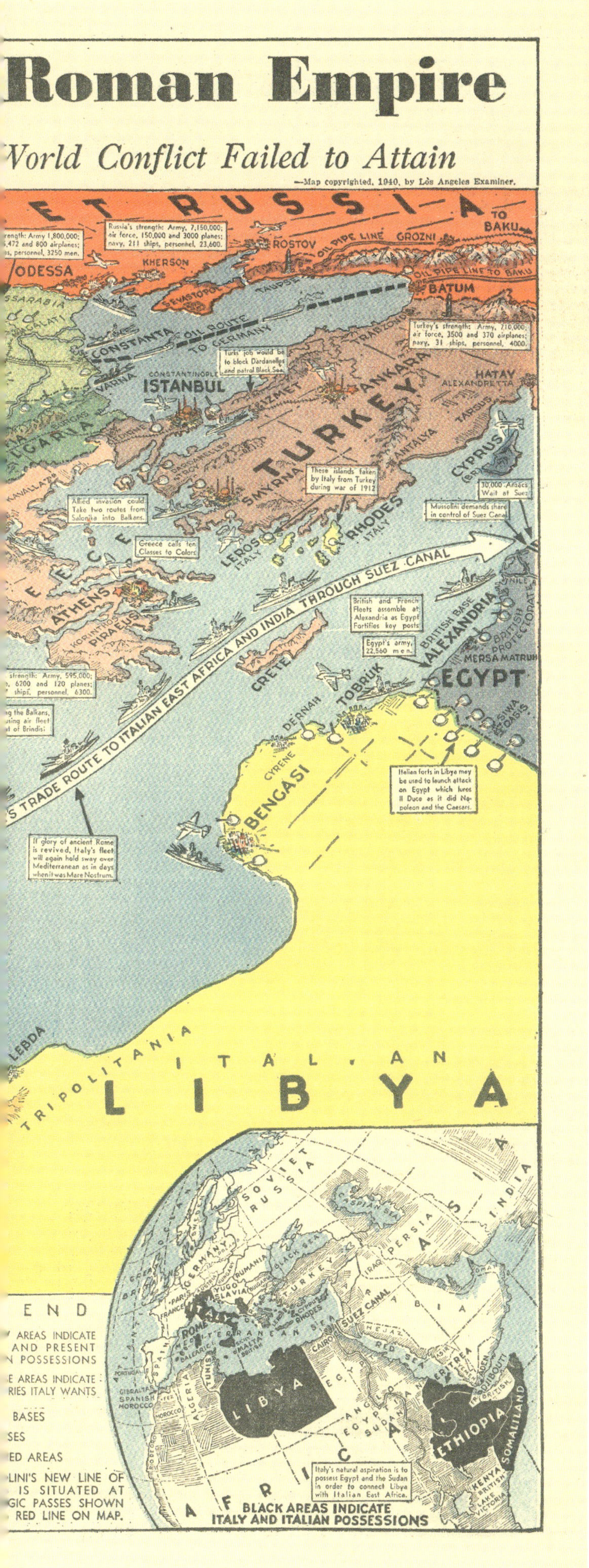

Burke (Howard): Italy Hopes War Will Bring New Roman Empire

Los Angeles Examiner, Los Angeles, 2 June 1940,
21 x 16 in. (53 x 41cm)

Published in the *Los Angeles Examiner*, this overview of Mussolini's overseas ambitions was printed as Italy was teetering on the very brink of war.

Newspaper magnate William Randolph Hearst, owner of the *Los Angeles Examiner,* was a fervent believer that America should keep out of any European war. He combined Anglophobia with unabashed admiration of both Adolph Hitler and Benito Mussolini. Both fascist dictators were even under contract to contribute a number of articles for Hearst publications.[132]

Hearst, throughout all his publications, was vehemently opposed to Franklin Roosevelt's policies, both domestic in the form of the New Deal and his foreign policy, in particular giving aid to Britain and the countries fighting fascism.[133]

In line with Hearst's editorial edicts this map presents a neutral if not actually sympathetic summary of Italy's war aims. Indeed Mussolini is usually referred to with the fascist honorific "*Il Duce*" throughout the map. The column of text in the lower left corner entitled *"What Italy Wants"* enumerates Mussolini's territorial, industrial and strategic objectives, along with speculation as to whether Italy would throw in her lot with the Allies as she had in the previous war or with Germany. It goes on to explain that *"Failure to get what she wanted from the First World War made Italy a coy and enigmatic neutral in this one – a Juliet to such Romeo nations as Germany, Britain and France. Whether their wooing will be her opportunity or her undoing remains to be seen as she approaches the brink of actual hostilities."*

[132] Somewhat unexpectedly, Mussolini's first article for a Hearst publication appeared in 1928 in the August issue of *Cosmopolitan Magazine* where he gave readers the benefit of his thoughts on the subject of "*Woman*", proclaiming amongst other insights that *"Her very nature imposes upon her a willing submission to the power and strength of the male."* Landers, James, 2010, *The Improbable First Century of Cosmopolitan Magazine*, Columbia, MO., p. 194. Hitler's first article appeared in September 1930 entitled "*Adolf Hitler's Own Story*" for which he was paid 1,000 marks (the equivalent of $240). This was considerably less than Mussolini was getting but sufficient, according to Putzi Hanfstaengl, Hitler's foreign press chief, to enable Hitler to stay in the plush Kaiserhof Hotel when he went to Berlin. See: Nasaw, David, 2013, *The Chief: William Randolph Hearst,* Boston, 2013.

[133] See Proctor, Ben, 2007, *William Randolph Hearst: The Later Years, 1911-1951*, Oxford.

It is striking that all the figures and statistics on this map treat France as a powerful combatant nation equal to her ally Britain. The rapidity of France's collapse in the face of German invasion was obviously inconceivable to the mapmakers. Only twelve days after this map was published, the Germans entered Paris. It was indeed the fear that France would capitulate before Italy even got involved that propelled Mussolini to declare war on Britain and France on 10 June 1940. He immediately launched an attack across the Alps with the aim of seizing territory in southern France before any armistice could come into effect.

The Battle of the Alps (10-25 June) was the last major engagement of the Battle of France. The French held defensive positions along a series of border fortifications called The Alpine Line or The Little Maginot. The French commander, General René Olry's seven divisions with numbers made up with elderly reservists, faced thirty Italian divisions backed up by the bombers of the Italian air force, the mighty Regia Aeronautica.[134] Despite the disparity of numbers the French managed to hold the Italians in check only conceding Menton before the declaration of the Armistice. In the general military collapse against the Germans elsewhere in France, this valiant defence in the south is often overlooked.

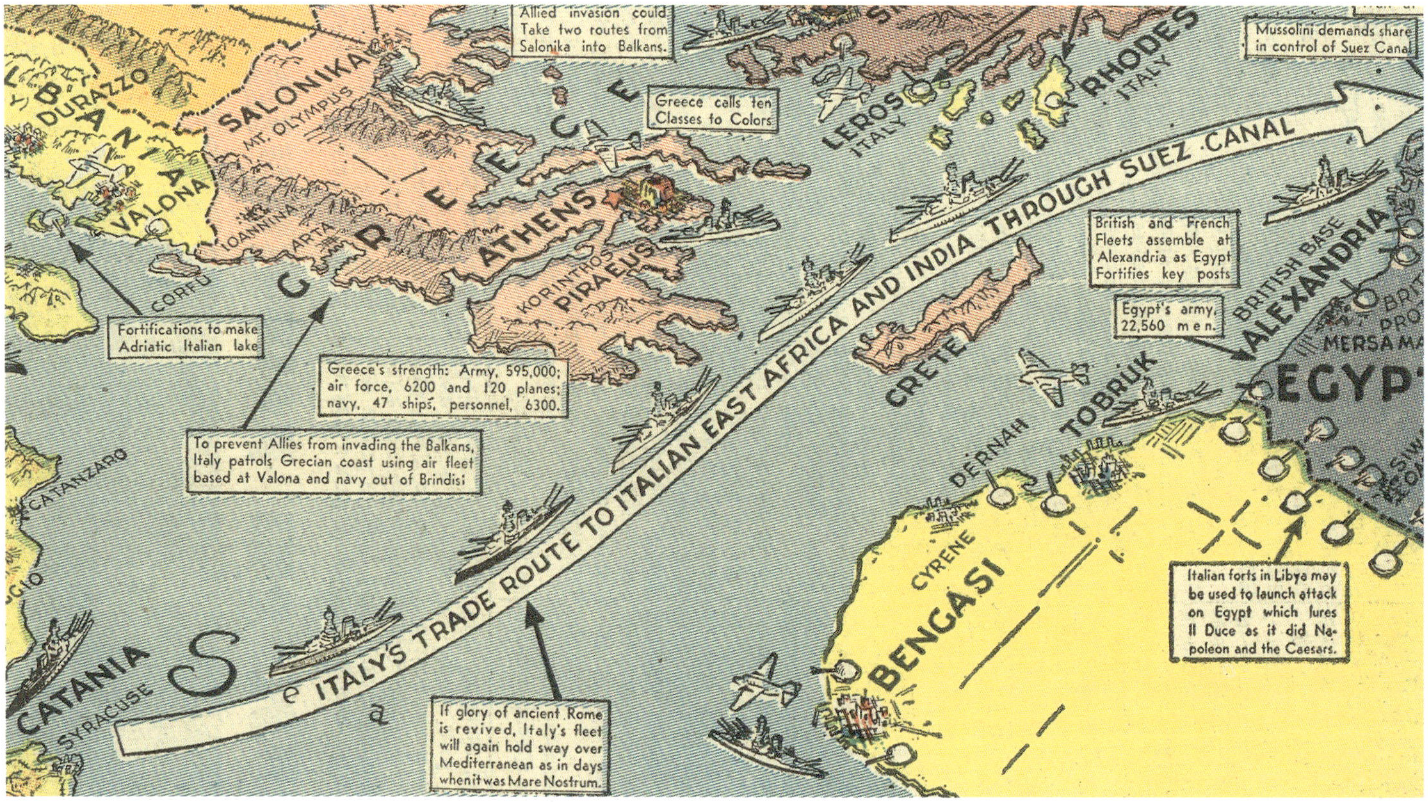

The predictions of Italy's intentions indicated on the map proved remarkably accurate. There is a note that *"Italian forts in Libya may be used to launch attack on Egypt which lures Il Duce as it did Napoleon and the Caesars"*. British Egypt and the Suez Canal were indeed irresistible to Mussolini and Italian forces under General Rudolfo Graziani invaded from Libya with 236,000 men in September 1940.[135]

[134] Italian divisions were 'binary' containing two regiments rather than the more usual three, and so were smaller than French divisions. The Italian soldiers were also often poorly led and equipped, and the foggy and mountainous conditions of Alpine warfare did much to negate the advantages of overwhelming numbers and air superiority. See: Knox, MacGregor, 1999 [1982], *Mussolini Unleashed, 1939-1941: Politics and Strategy in Fascist Italy's Last War*, Cambridge.

[135] Matanle, Ivor, 1995, *World War II*, Brookvale, N.S.W., p. 75.

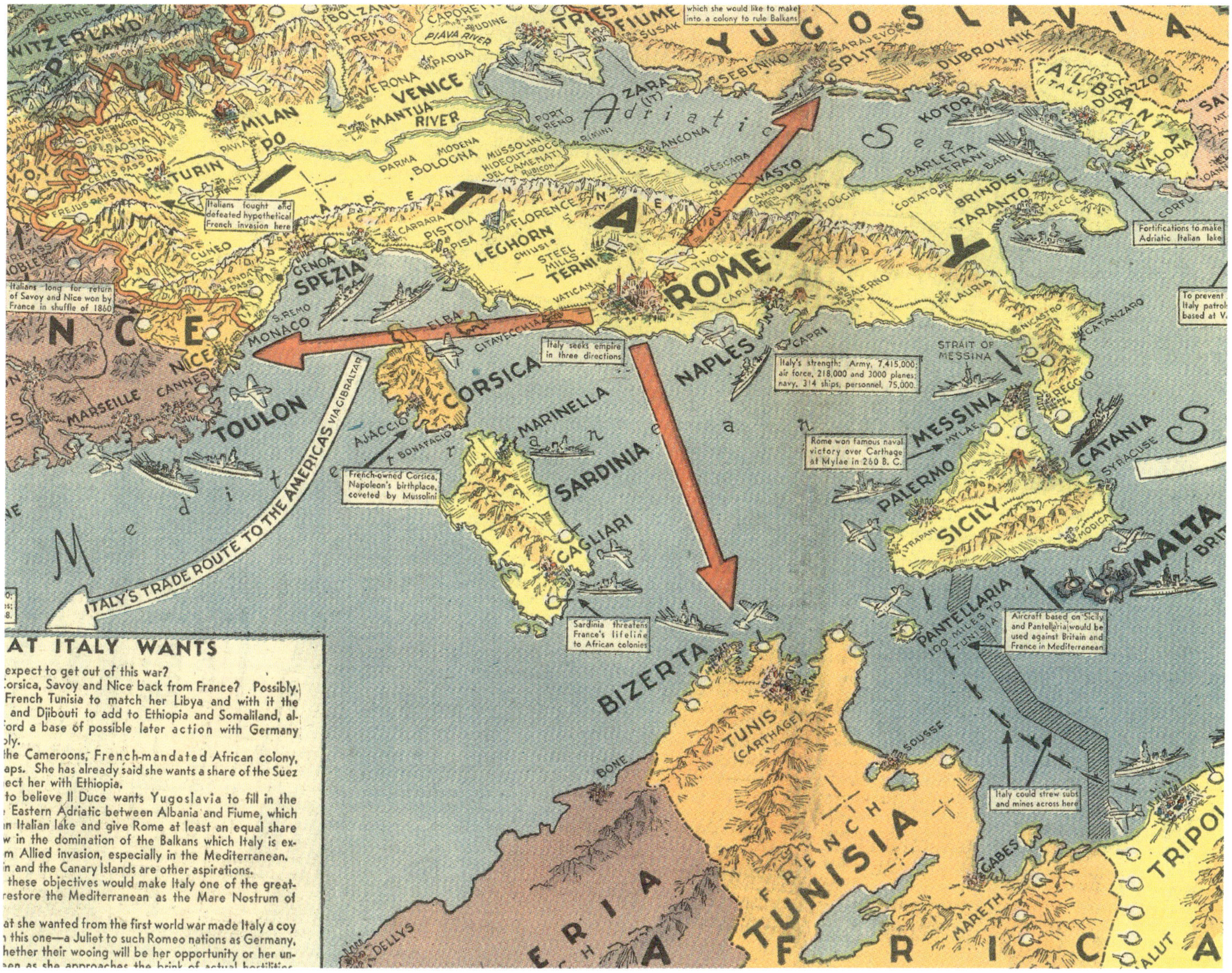

Italy invaded Yugoslavia alongside Germany in April 1941 and after the cessation of Vichy France in 1942, she would go on to occupy the south of France and Corsica, all designated as Italian targets on the map.[136]

[136] Tomasevich, Jozo, 2002, *War and Revolution in Yugoslavia, 1941-1945: Occupation and Collaboration*, Stanford, CA., p. 46.

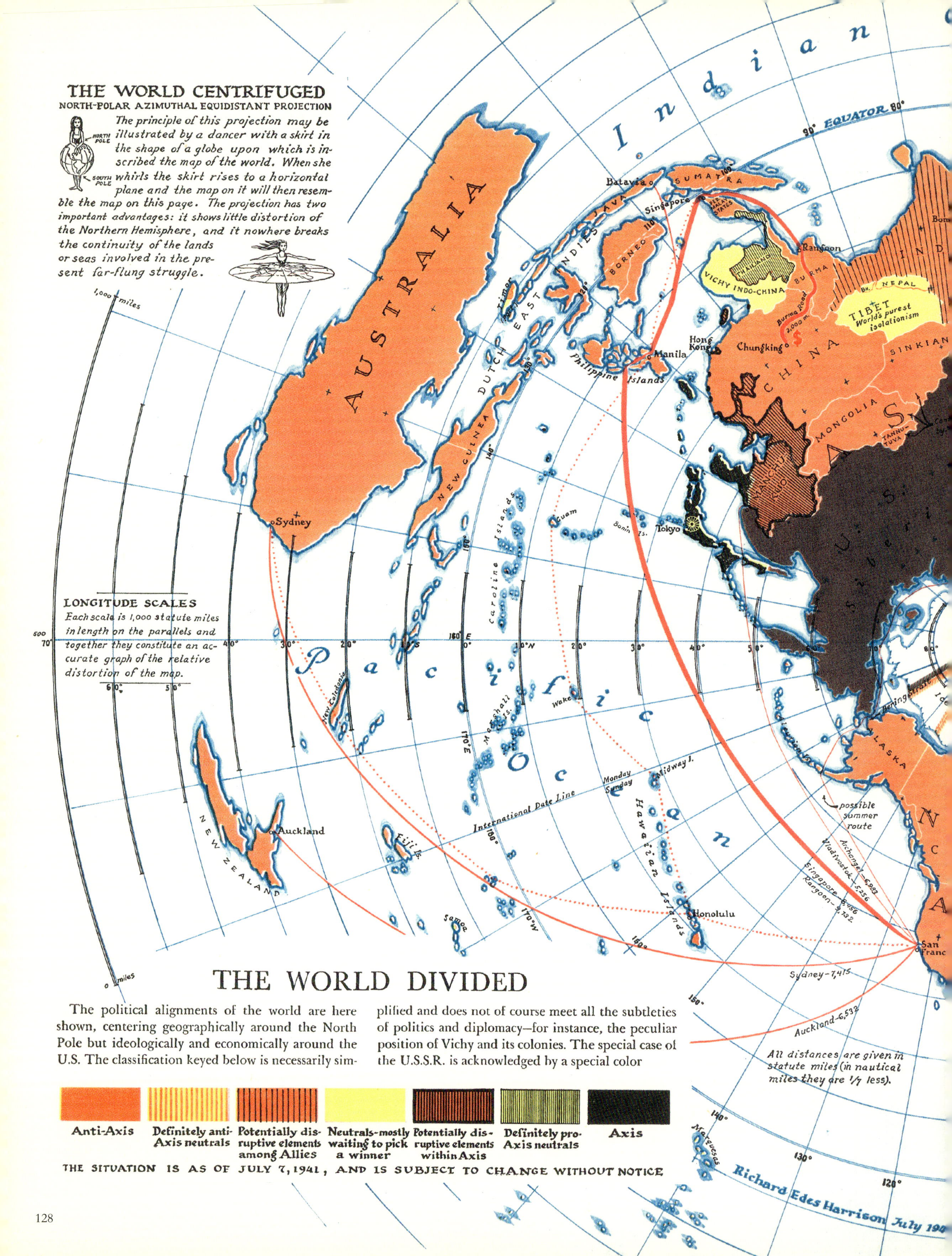

THE WORLD DIVIDED

The political alignments of the world are here shown, centering geographically around the North Pole but ideologically and economically around the U.S. The classification keyed below is necessarily simplified and does not of course meet all the subtleties of politics and diplomacy—for instance, the peculiar position of Vichy and its colonies. The special case of the U.S.S.R. is acknowledged by a special color

Anti-Axis	Definitely anti-Axis neutrals	Potentially disruptive elements among Allies	Neutrals-mostly waiting to pick a winner	Potentially disruptive elements within Axis	Definitely pro-Axis neutrals	Axis

THE SITUATION IS AS OF JULY 7, 1941, AND IS SUBJECT TO CHANGE WITHOUT NOTICE

THIS is a map of a world struggle that will mark a turning point in civilization. The struggle divides the world into two huge camps: those who are for us (in various degrees) and those who are against us (likewise in various degrees). Between these extremes there are the neutrals, some leaning one way, some another. All these gradations are explained in the key on the opposite page.

Several important points regarding this map should be noted. The first is that the entire conflict pivots around the U.S. This is not due merely to the placing of the U.S. in the center of the spread: it lies, rather, in the very nature of the struggle. The U.S. is the pivot *(a)* because of its geographical position, which is central to the oceans; *(b)* because of its ideological position, which makes it the hope of all free peoples; and *(c)* because of its policy of supplying the enemies of the Axis with the materials of war, via the lend-lease lines.

Because the U.S. is pivotal, and because the U.S. has not yet taken military action, the map is political rather than strategical. Yet there are several strategical considerations:

A. The Western Hemisphere lies between the shores of the great land mass of the Eastern Hemisphere. If those shores were controlled by the enemy, the U.S. would find itself in the grip of a gigantic pincers. At present there are two bastions preventing this—and only two. One is Britain, fed by the big lend-lease line from the East Coast. The other is China, fed by a thinner lend-lease line to the Burma Road.

B. The enormous importance of the U.S.S.R. shows up on this map. If Hitler can crush the Russian Army and set up a puppet government, this will constitute a direct territorial threat to the U.S. The thin red line running from San Francisco around the polar cap to Archangel is quite academic as a lend-lease route, but it serves to show how close we are to that theatre of action. Air attack from this quarter is not beyond possibility. And if the Soviet Union were to fall, Hitler would be, mainland to mainland, within fifty-seven miles of U.S. territory across the Bering Strait.

As a result of its pivotal position the U.S. is fighting the war all over the world. It is intimately interested in the affairs of almost every nation outside the black area of the Axis. The lend-lease lines show the lines of major involvement, but there are also such problems as Eire, India, and the Union of South Africa. If Hitler won the Irish Government to his side the repercussions in the U.S. would be tremendous. The Union of South Africa is really on our side, but internally divided. And India, which Hitler may be able to reach through the U.S.S.R., wants freedom now. Finally, there is Latin America. We have sought in many ways to bind the Latin-American countries to our cause—by dollar loans, by military missions, and by education and propaganda. The results so far have been anything but satisfactory.

Such is the world we face. Largely, it is the result of our own failure to be realistic in the past. Now we have to fix it. And our first move had better be to become realistic for the future.

Lend-lease water routes

Lend-lease air routes

U.S. military missions

U.S. financial aid

Harrison (Richard Edes): The World Divided - The World Centrifuged

lithographed map, Fortune Magazine, New York, 1941, 20 x 14 in. (51 x 36 cm)

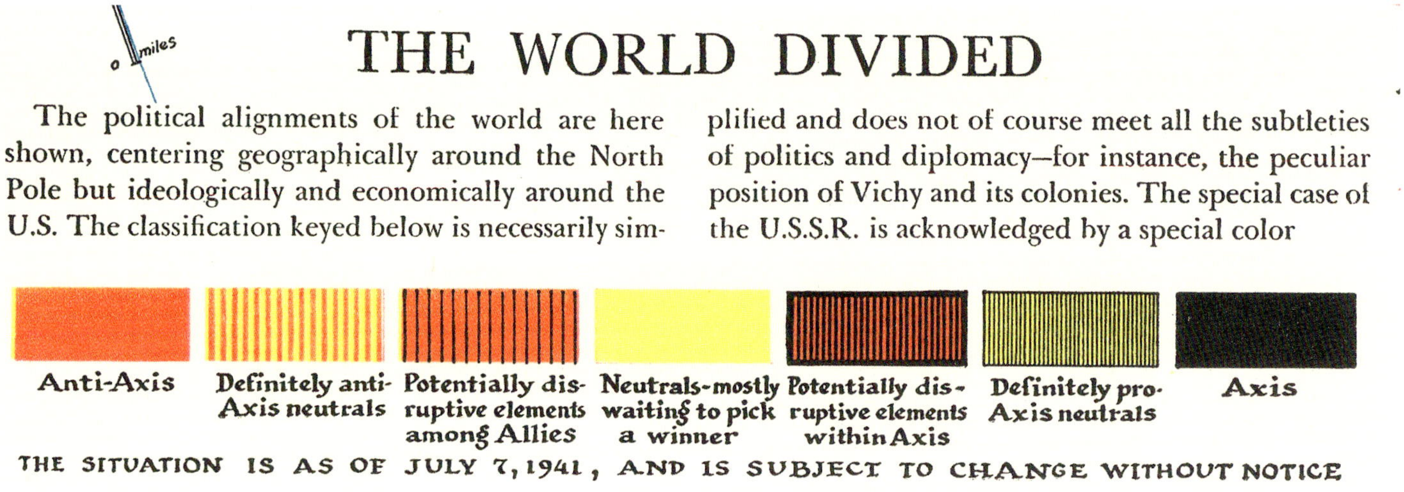

This is an outstanding graphic summation of the political allegiances divided between the Axis and the Allied powers, drawn up while America was still militarily uncommitted.

It was on this map that Harrison most fully realised his idea that it was only a polar projection centred on the North Pole that revealed the geographical positions of the combatant countries and their relationships to the U.S. Unlike the impression of isolation conferred on the U.S. by world maps on Mercator's Projection, this polar view emphasises the closeness of the conflict to North America.

Despite their nominal neutrality in July 1941, this map emphatically aligns the U.S. with the anti-Axis forces. The world is divided and colour-coded relative to the local adherence to one side or the other.

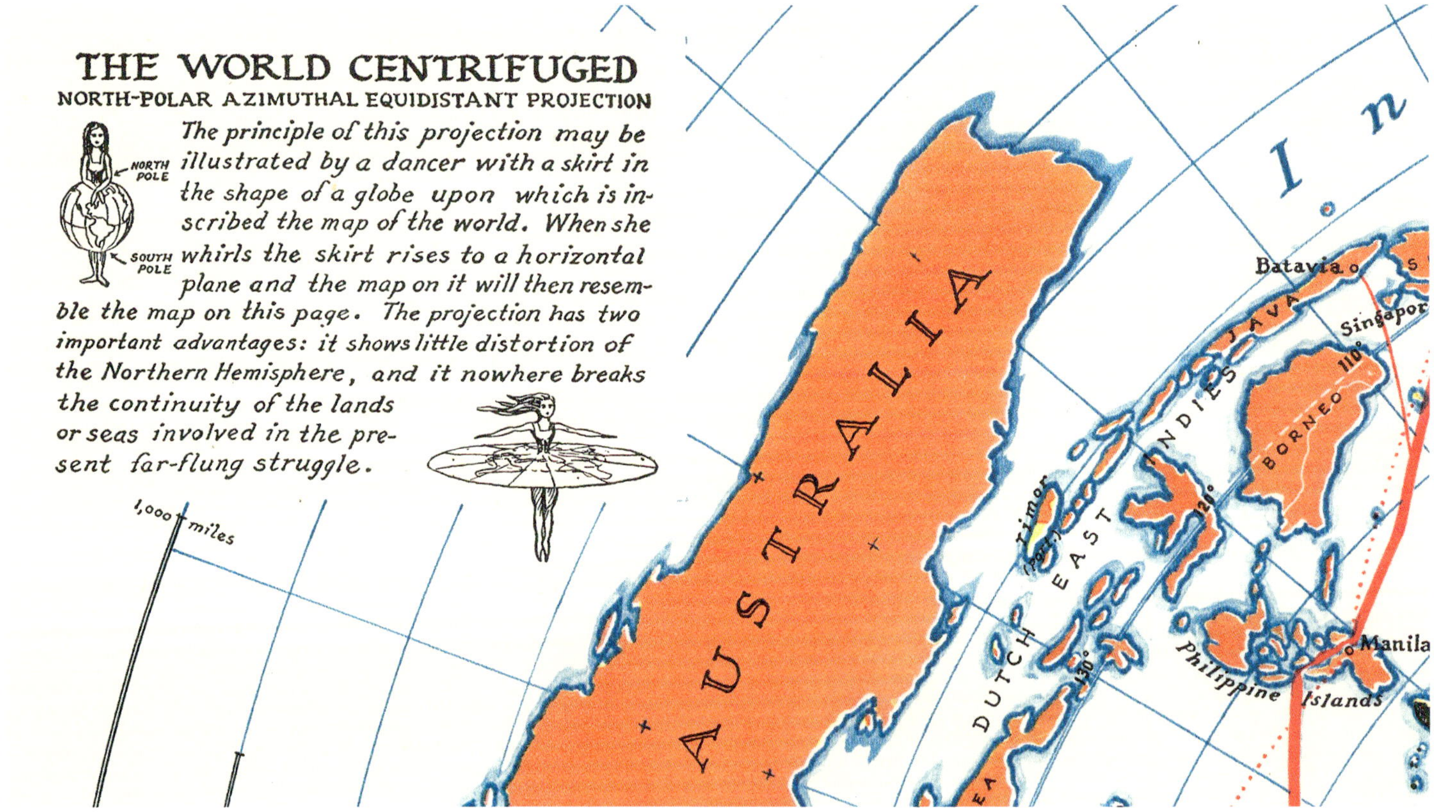

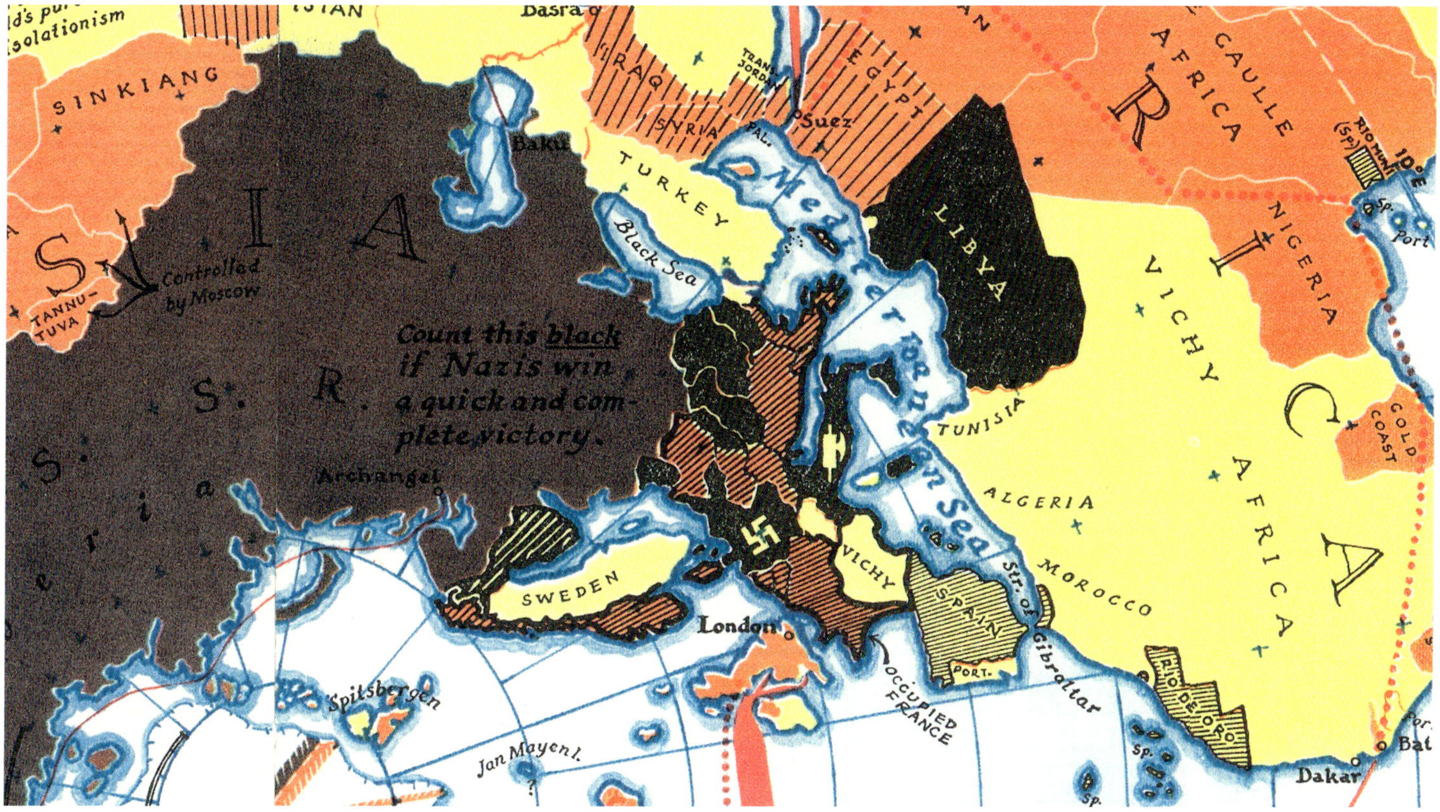

Most significant for the purposes of this map is the uncertain position of Russia. By 7 July 1941, Operation Barbarossa, the German invasion of Russia, had only been underway for a couple of weeks but already the armoured columns of the Wehrmacht were deep inside Russia sweeping all before them. The map warns that if Russia were to fall this would put the Axis forces in Siberia only fifty-seven miles from U.S. territory. It further cautions that in the event of an Axis victory in the Eastern Hemisphere, North America would "find itself in the grip of two gigantic pincers".

There are very different messages coming from the publications owned by Henry Luce, such as *Time* and *Fortune* magazines, and those issued by the Hearst newspapers. Here Harrison stresses the danger of allowing the Axis forces to prevail while Howard Burke's map [see map on p.106] for Hearst's *Los Angeles Examiner* warns of the danger to the U.S. if she should join the Allies and be defeated.

The twentieth-century saw the rise of geopolitics as a respected academic discipline and as an increasingly valuable method of interpreting events. Both in Europe and the U.S. artists found a shared belief in maps being "the Chief tool of the geographer and layman alike in visualizing the geography of World War II".[137]

"The main purpose [of global perspective maps] is not to locate supply routes and battle lines, [...but rather] to show why Americans are fighting in strange places and why trade follows its various routes. They emphasize the geographical basis of world strategy." [138]

[137] Harrison, Richard Edes, 1943, The war of the maps, *Saturday Review of Literature*, New York, p. 25.

[138] Harrison, Richard Edes, 1944, *Look at the world: The Fortune world atlas for world strategy*, New York, p. 1.

C. S. Hammond & Co., Inc.: Defence Map of United States showing the Mobilization of our Military and Economic Sources

C. S. Hammond & Co., Inc.: Defence Map of United States showing the Mobilization of our Military and Economic Sources

lithographed map, C. S. Hammond & Co., New York, 1941,
31 x 21 in. (79 x 53 cm)

Published prior to United States' entry into the Second World War, this is a morale-boosting tribute to American economic power and the strength of their armed forces.

The map itself shows the U.S. divided into Army Corps areas. Marked by red and blue symbols are the forts, airfields and stations of the infantry, air force and navy, while strategic industrial and production facilities are shown with grey symbols. These include oil wells, chemical plants, mines, and steel and aluminium facilities. The majority of military bases are concentrated along the north-east coast, with the west coast limited to three clusters around Glendale, San Francisco and Seattle.

The map is surrounded by a border incorporating many of the leading war machines of land, sea and air. Prominent in the foreground is the Grumman fighter and a Consolidated B-24 *Liberator.* Although aircrews generally preferred the more robust qualities of the better-known Boeing B-17 *Flying Fortress*, the *Liberator* with its greater speed and range became the most produced heavy bomber of the war.[139]

The air force was naturally the youngest arm of the military and this is reflected in the style of the squadrons' insignia. Unlike the staid unit badges of the army and navy, its youthful exuberance is reflected in its insignia often incorporating cartoon figures.

Caleb Stillson Hammond started C.S. Hammond & Co. in Brooklyn, New York, in 1900, having previously been head of Rand McNally's New York office. The company would stay under family control until 1999 when it was bought by Langenscheidt Publishers Inc. The Second World War saw an increase in the production of their pictorial maps which helped propel the company into a position just behind Rand McNally during the 1950s and 1960s. In December 2002 the extensive archives of Hammond World Atlas Corporation was acquired by the Library of Congress.[140]

[139] Burtness, Robert A. 2012, *The Santa Barbera B-24 Disasters: A Chain of Tragedies Across Air, Land & Sea*, Charleston, SC., p. 13. Over 18,400 were built by the end of the war.

[140] www.loc.gov/loc/lcib/0305/hammond.html

Guenther (Lambert): Safeguarding our American Liberty

Guenther (Lambert): Safeguarding our American Liberty

lithographed map, C. S. Hammond & Co., New York, 1941, 21 x 31 in. (53 x 79 cm)

Similar to the Hammond *Defence Map of the United States* and published the same year, this map once more extols the strength and the global reach of the U.S. military might. The artist, Lambert Guenther, has overlaid a geographical world map with a view of the American air force and navy in the air and at sea surmounted by the Stars and Stripes.

The map notes the number of current warships together with those in production, emphasising the massive build-up of American naval force in preparation for the increasingly inevitable Pacific-wide war. Another seventeen battleships are planned to join the existing fifteen while cruisers will be increased from thirty-seven to eighty-five. Significantly the most important vessel, the aircraft carrier, is being increased from six to eighteen. The surprise attack by the Japanese on Hawaii would drastically change the balance of this naval construction. The devastating triumph of Admiral Yamamoto's carrier-borne air force over the unprepared U.S. Pacific fleet and the sinking of four of the eight battleships present at Pearl Harbor, together with the fortuitous escape of the two aircraft carriers effectively imposed the ultimately successful carrier based strategy on the Americans.[141]

This map also draws attention to the other crucial theatre of war in the coming conflict – that of the air. The maximum range of the fleet and plane patrolling range is shown surrounding North America and the Pacific bases centred on Pearl Harbor.

The artist, Lambert Guenther, was born in Sweden in 1909, settled in New York and established himself as an artist principally working on paintings and murals.

[141] Paradoxically, despite the overwhelming success of the carrier attack on Hawaii, a significant part of the Japanese naval high command still believed that the naval war would ultimately be won in a great capital ship battle. They carefully husbanded their battleships which included the *Yamoto* and the *Musashi* which, at 72,000 tons displacement and 800 feet length, were the largest battleships ever built. When these colossal ships finally ventured outside the safety of home waters, they were both sunk by American carrier-borne dive bombers and torpedo planes. Beevor, Antony, *The Second World War*, London, 2012, p. 629.

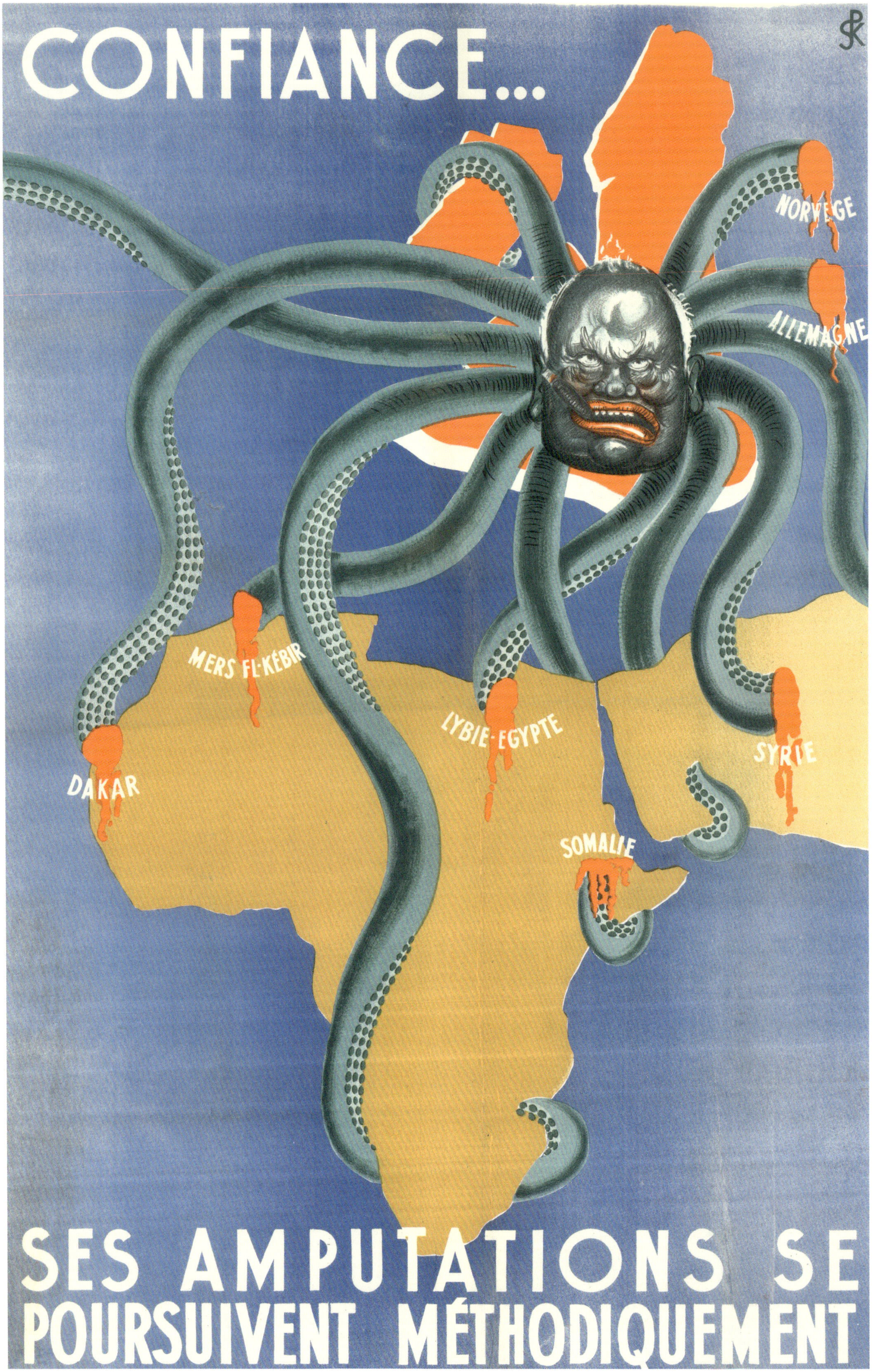

S. P. K.: Confiance – Ses Amputations se Poursuivent Methodiquement

S. P. K.: Confiance – Ses Amputations se Poursuivent Methodiquement [Be Reassured – The Amputations are Proceeding Methodically]

lithographed map, [Paris?], 1941, 32 x 47 in. (81 x 120 cm)

This potent propaganda map, designed to inspire revulsion and hatred of the English among the French, while frequently described as a 'Vichy' map is likely to have been printed in Paris by the Propaganda-Abteilung Frankreich, the German propaganda unit in occupied France.

A grotesque octopus, with the face of Winston Churchill, is shown attempting to seize various French colonial possessions. The demonic Churchill, eyes burning with fury, has the full red lips seen in many anti-Semitic caricatures of the period, popularised by such magazines as *Der Stürmer*. The tentacles of the octopus have been severed in seven places, all of which purport to show British defeats during the first two years of the war. The most significant locations for its French audience would have been Dakar and Mers-el-Kébir in North Africa and its former colonies in the Levant.[142]

The defeat of France in 1940 and the subsequent armistice with the Germans had left many within France and its colonies with deeply conflicted loyalties. While some supported the Free French under Charles de Gaulle, a significant number of Frenchmen felt their allegiance lay with the "legitimate" government of Marshall Pétain in Vichy France.[143] This was mirrored across French North Africa, West Africa and Indochina.

On 3 July 1940, the first major confrontation between the British and Vichy French forces occurred in the North African port of Mers-el-Kébir, where the largest concentration of French naval strength lay at anchor including the *Strasbourg* and the *Dunquerque,* two of the most powerful battleships in the world. A strong British force under the command of Admiral James Somerville, fearful that the French fleet would eventually fall into the hands of the Nazis, delivered an ultimatum. The French ships were given the choice of joining forces with their former allies, to be escorted with reduced crews to the French West Indies or America to wait out the war or finally to be scuttled by their crews. Somerville informed his opposite number, Admiral Marcel-Bruno Gensoul, that in the event of non-compliance the British fleet would fire upon the French fleet.[144] At 17:54, after several hours of inconclusive negotiations and on the direct orders of Churchill, the British opened fire at point blank range. In some thirty salvos, the British sank six ships including the battleship *Bretagne* and killed 1,297 Frenchmen.[145] This ruthless and still contentious episode caused a rift in Anglo-French relations, but at the same time it confirmed, especially to a sceptical United States, Britain's determination to do whatever was necessary to defeat Nazi Germany.[146]

[142] Barber, Peter & Harper, Tom, 2010, *Magnificent Maps*, London, p. 165.

[143] It should be noted that the legitimacy of the Vichy government was also recognised by Britain until the French themselves broke off diplomatic relations after the sinking of the French fleet at Mers-el-Kebir in July 1940. After the French capitulation, the Americans moved their embassy from Paris to Vichy and continued full diplomatic relations until the Allied invasion of French North Africa in November 1942. Even the Russians maintained full diplomatic relations until the Vichy regime expressed support for the German invasion of Russia.

[144] Beevor, Antony, 2012, *The Second World War*, London, p. 124.

[145] 'the orders of His Majesty's Government': The National Archives, Kew, *ADM 399/192*

[146] 'tantamount to a declaration of war': The National Archives, Kew, *ADM 199/391*; Beevor, Antony, 2012, *The Second World War*, London, p. 125.

The Battle of Dakar in Senegal in September 1940 was the first time the forces of the Free French directly faced those of Vichy France. Dakar was important because it held the gold reserves of both the Bank of France and the Polish Government in exile, and also its harbour sheltered remnants of the French fleet. Charles de Gaulle, leader of the Free French, was convinced that he could persuade the Vichy forces to join the Allied side in an important propaganda victory. In the event, the Vichy troops refused to negotiate and fired upon the Allies. The British, supported by Australian and Free French ships returned fire and on 7 July 1940 attacked the city.[147] De Gaulle's distaste for Frenchmen killing Frenchmen caused him to call off the assault. After several inconclusive actions and some damage to ships on both sides, the Allies withdrew leaving the Vichy forces in control of the port.[148] The Battle of Dakar, while strategically inconsequential severely damaged de Gaulle's reputation and it is plain why the propagandists would want to feature it on the map.

The severed tentacle in Syria actually represents a major defeat for Vichy during the Syria-Lebanon campaign. By July 1941 the Vichy forces in Syria had been comprehensively defeated by a mixed contingent of British and Free French. De Gaulle was deeply disappointed that of the 38,000 French PoWs captured during the campaign, 32,000 when given the choice of joining the Free French and fighting against their occupiers or being repatriated, elected to return to an enemy-dominated France.[149]

Propaganda pieces produced by defeated or discredited regimes rarely survive the downfall of their creators. The unattractive consequences of being found after 1944 with such a vicious caricature of Winston Churchill with its associated taint of collaborationism makes this one of the rarest of pictorial maps.

[147] Marder, Arthur J. 1976, *Operation 'Menace': the Dakar expedition and the Dudley North Affair*, Oxford.

[148] Beevor, Antony, 2012, *The Second World War*, London, p. 174.

[149] Mollo, Andrew, 1981, *The Armed Forces of World War II*, New York.

NORVEGE
ALLEMAGNE

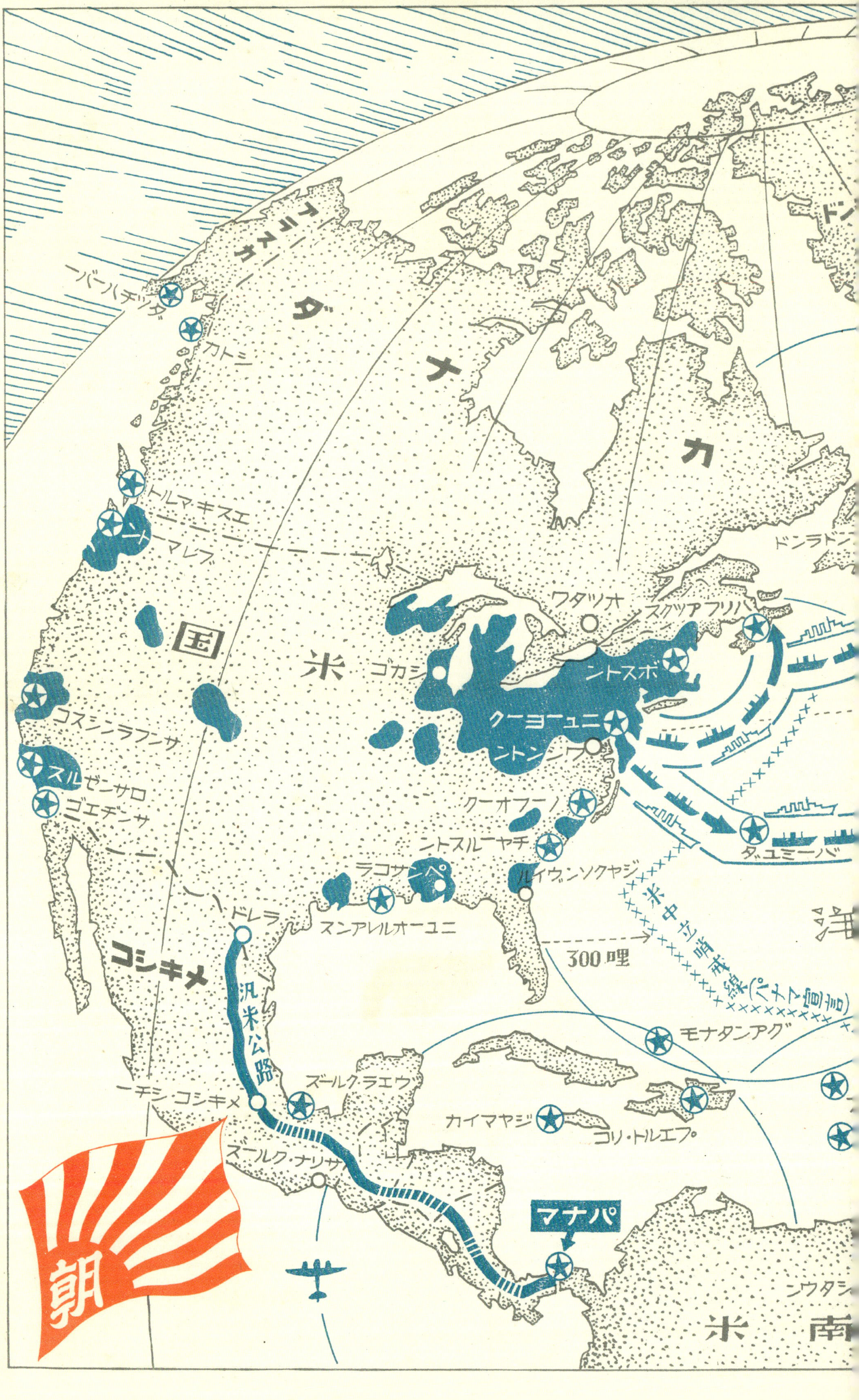

アラスカ
カナダ
米国
メキシコ
ダッチハーバー
シトカ
エスキマルト
ブレマートン
サンフランシスコ
ロサンゼルス
サンチエゴ
シカゴ
オタワ
ハリファックス
ボストン
ニューヨーク
ワシントン
ノーフォーク
チャールストン
ジャクソンヴィル
ペンサコラ
ニューオルレアンス
ラレド
汎米公路
メキシコ・シチー
ヴェラクルーズ
サリナクルーズ
パナマ
グアンタナモ
ジャマイカ
プエルト・リコ
バーミューダ
300哩
米中立哨戒線(パナマ宣言)
朝
南米

危機の大西洋要圖 米國參戰の爆發點

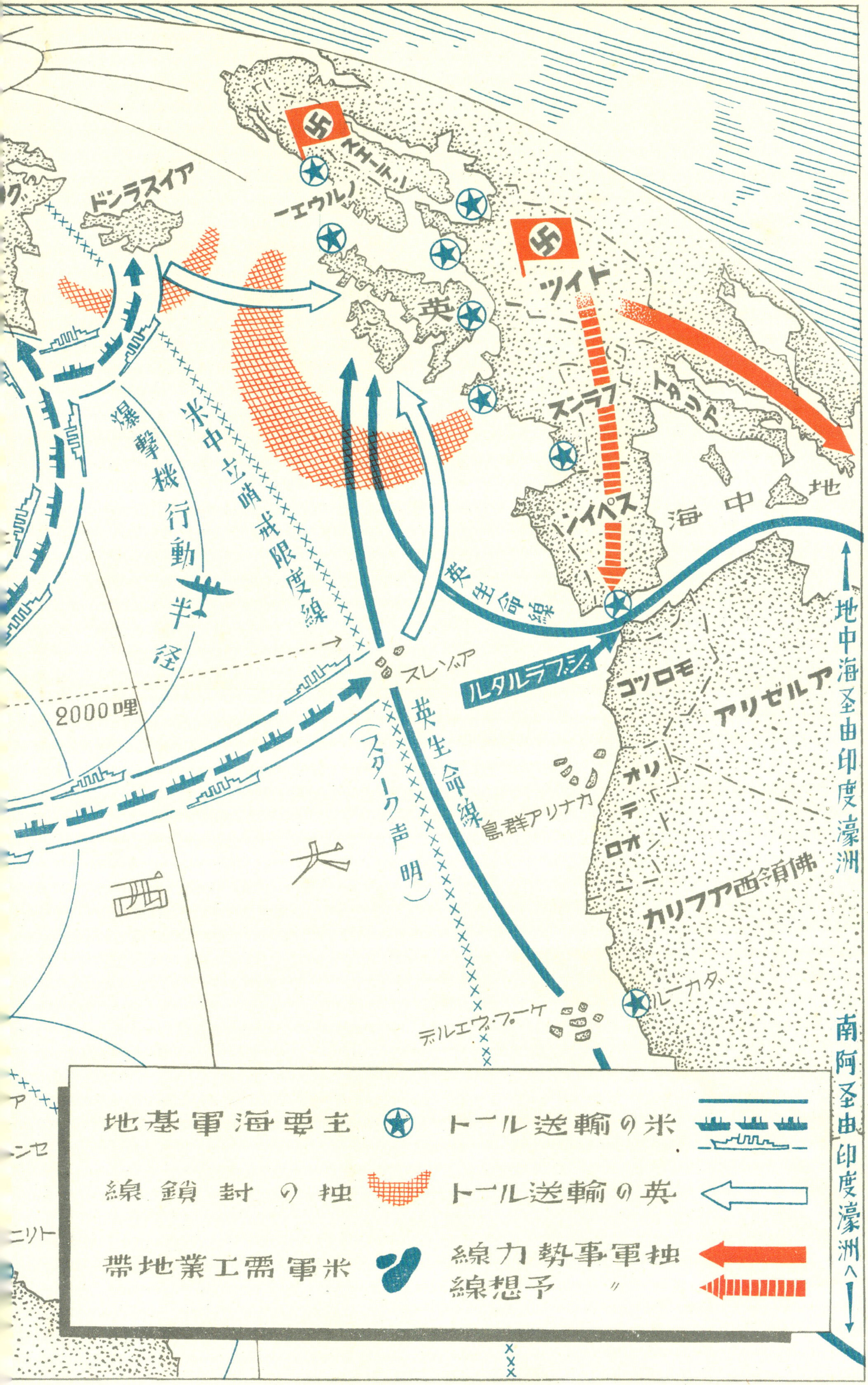

Anon: 危機の大西洋要図 米国参戦の爆発点 [Rough Sketch of the Atlantic in a critical situation – Explosion points in the U.S. participation of the war]

Anon: 危機の大西洋要図 米国参戦の爆発点 [Rough Sketch of the Atlantic in a critical situation – Explosion points in the U.S. participation of the war]

lithographed map, Japan, 1942, 29 x 20 in. (78 x 54 cm)

This is an intriguing Japanese view of the situation and events in Europe and United States and in the Atlantic Theatre of War.

There was surprisingly little cooperation or coordinated action between the Axis partners of Germany and Japan. Germany wanted Japan to attack their mutual enemy the Soviet Union in Siberia to relieve pressure on their Eastern Front, while the Japanese were determined to push on into China and the resource rich regions of the Pacific. Despite this the Japanese were aware that events in the Atlantic, and how the United States' vast effort to sustain the Allies' war aims in Europe and North Africa directly impacted on the war in the Far East.

The principal convoy routes are emphasised on the map. The northern route via Greenland and Iceland relied on the protection of air cover, while safety on the southern route was dependent on the vastness of the ocean. The areas of maximum danger from the U-boat wolf packs are indicated by red-hatched areas of the sea.

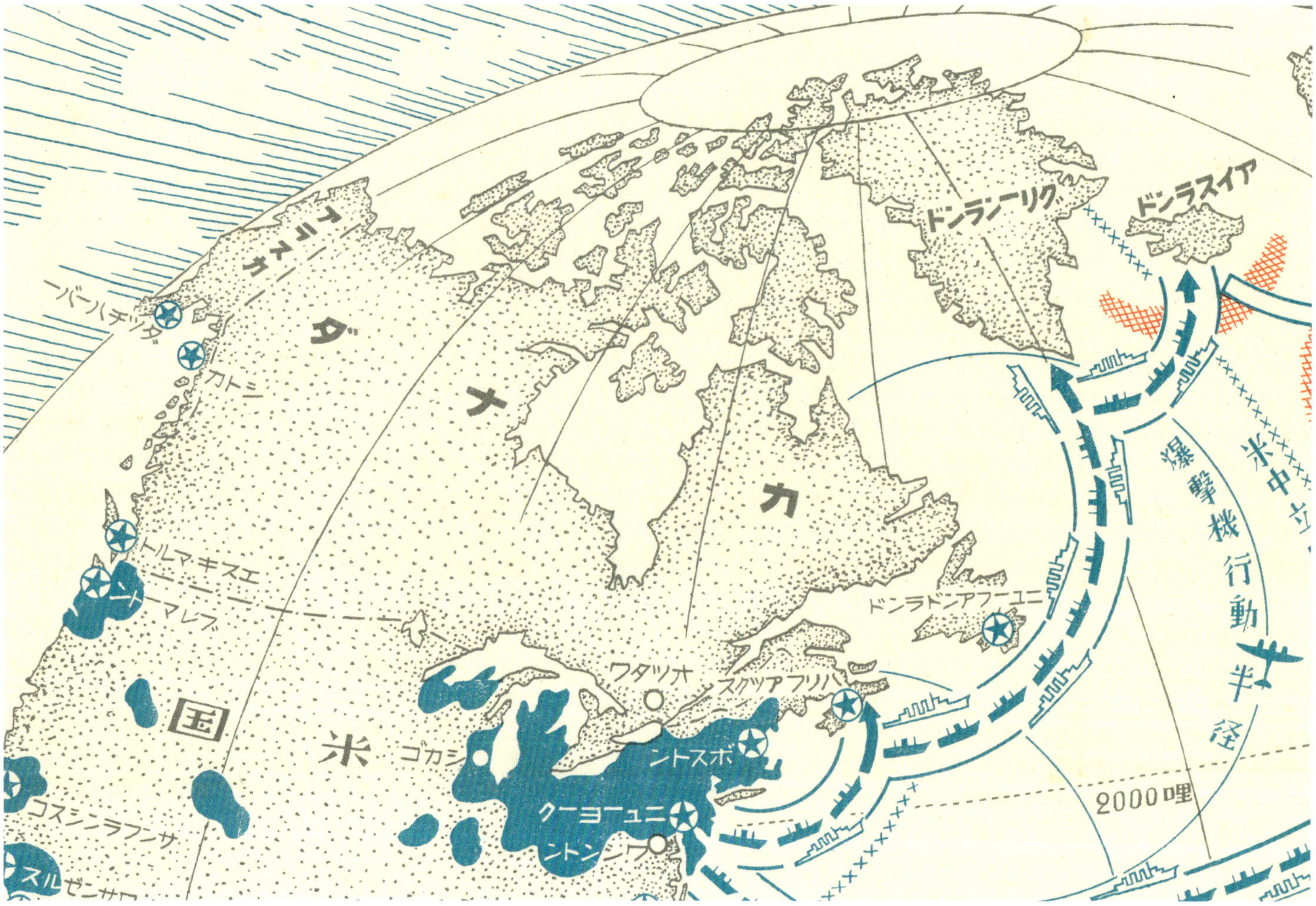

An interesting feature of this map is how the mapmaker, whose perception has perhaps been distorted by the Japanese tactical preoccupation with maritime choke-points, has incorrectly predicted the future actions of both America and its own ally, Germany.

Firstly the Japanese have indicated with a broken line of advance that they anticipate the Americans marching down through Mexico and Central America to seize the all-important Panama Canal by force. Conversely on the other side of the Atlantic, with a red broken line marked *"Future campaigns"*, Germany is shown advancing through neutral Spain in order to take control of Gibraltar.

The flag in the lower left-hand corner is that of the Asahi Shimbun Company, also known as Asahi Press, the publisher of this map. Established in 1879, Asahi Shimbun is one of Japan's oldest newspapers. It had and has a generally liberal viewpoint which brought it into conflict with both the Imperial government and indeed its own controversial editor-in-chief, Taketora Ogata. Ogata was a leading member of the Genyōsha, an ultranationalist group of organised crime figures and far-right politicians, formed in 1881. In November 1945, Asahi Shimbun's president and the entire executive staff resigned en masse for having compromised the paper's principles during the war.[150]

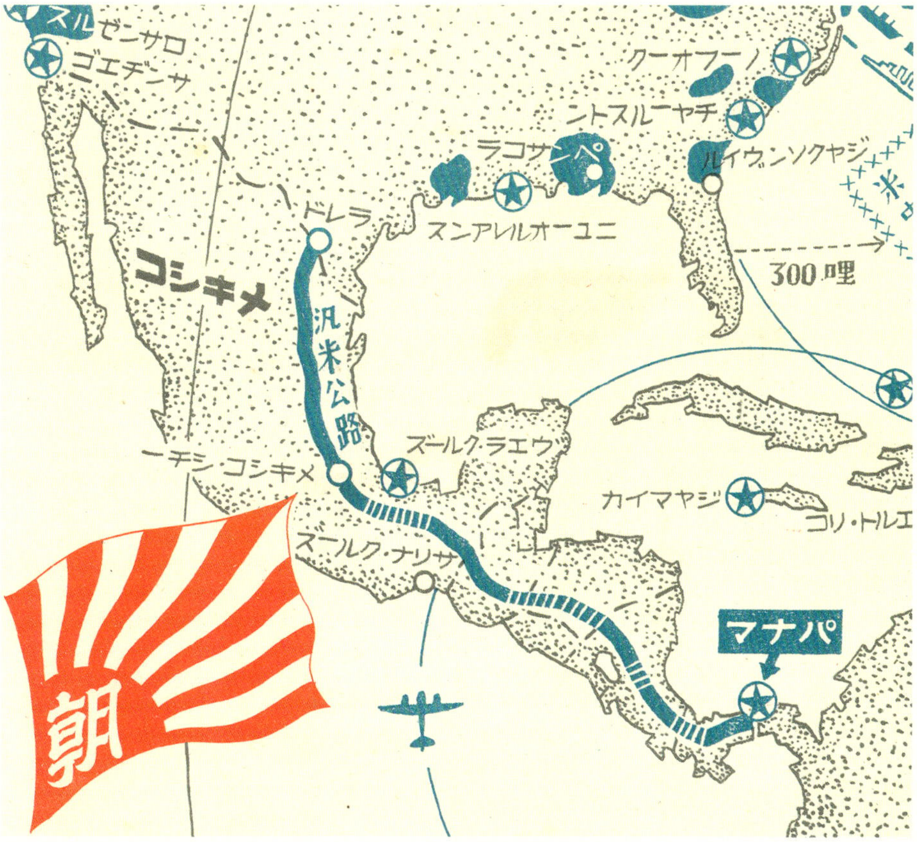

[150] Asahi Shimbun Shashi, *Asahi Shimbun Sha, 1990–1995*, Tokyo and Osaka; Kodansha, 1983, *Kodansha Encyclopedia of Japan*, Tokyo and New York; Merrill, John C. and Harold A. Fisher, 1980, *The world's great dailies: profiles of fifty newspapers*, pp. 59-67.

JEAN FORT

Fort (Jean): [Africa]

Fort (Jean): [Africa]

lithographed map, Bedos & Cie. Imprimeurs, Paris, 1942, 46 x 30 in. (116 x 77 cm)

Similar in style to the Churchill Octopus map, this powerful image of Roosevelt and Churchill in a tug-of-war over Africa was created by Jean Fort in 1941 for PAF (Propaganda-Abteilung Frankreich), the German propaganda unit in occupied France.

The poster was designed to inflame French patriotic feeling and resentment after the Allied invasion of French North Africa at the start of Operation Torch. Vichy France had over 105,000 men stationed in North Africa, and it was a matter of vital strategic concern to both the Allied and the Axis powers whether they would resist the foreign invaders or join their compatriots in the Free French.

The French forces certainly had ambivalent feelings, at the very least, towards the Allies. Some like Admiral François Darlan, who became commander-in-chief in North Africa, believed that the Germans would ultimately win the war and that France's only chance to rise again would be to remain neutral. Others regarded Marshal Philippe Petain's government, however constrained and compromised, as the legitimate government of France to which they'd pledged allegiance. This was often reinforced by a deep dislike of the British, regarded by many Frenchmen as having deserted their allies at Dunkirk, and which was intensified by the destruction of the French fleet at Mers-el-Kebir. How much of this was exacerbated by the relentless dissemination of pro-German, anti-Allied propaganda is impossible to judge, but it is certainly true that before 1944 there were many more Frenchmen in uniform fighting against the Allies than there were supporting them in the Free French.[151]

The Wehrmacht Propaganda-Abteilung, WPr. (Wehrmacht Propaganda Department) was the first of its kind in military history. Before its establishment, no military department had existed whose activities embraced all areas of modern propaganda and functioned as an independent branch within the armed forces. The first initiatives to set it up were in 1935-1936, principally by the Ministry of Propaganda. There was a general consensus within the Wehrmacht command that the failure of German propaganda in the First World War had a major impact on the collapse of the German rear and had led to the German defeat.[152]

The Propaganda-Abteilung Frankreich (PAF) of the Wehrmacht High Command in France implemented active propaganda across the Soviet Union, the United States, Europe and the Middle East throughout the Second World War. Based in the Hôtel Majestic in Paris, it had a budget of over 90 million francs a month and employed over 1,000 personnel at its height.[153]

[151] The complex situation for French servicemen can be illustrated by the war record of Pierre le Gloan, France's fourth highest ranking Second World War air ace. His confirmed tally of eighteen kills consisted of four German and seven Italian planes during the *Battle of France* and seven British planes during the Syrian Campaign. He was once more fighting in the Free French on the side of the Allies when he was killed while crash-landing on 11 September 1943.

[152] Buchbender, Ortwin, 1978, *Das tönende Erz: Deutsche Propaganda gegen die Rote Armee im Zweiten Weltkrieg*, Stuttgart, pp. 16-19.

[153] Dorléac, Laurence Bertrand, 2008, *Art of the Defeat: France 1940-1944*, Getty Publications, pp. 45-46.

Manche (Louis Emile): De Dollarpoliep [The Dollar Octopus]

lithographed map, Afd. Propaganda van het Departement van Volksvoorlichting en Kunsten, Tijdsduur, Holland, 1942, 23 x 17 in. (59 x 43 cm)

Manche (Louis Emile): De Dollarpoliep

The propagandists' favourite image of a grasping octopus is used here to represent U.S.'s use of its financial muscle to extend its territory and influence around the globe.

The tentacles have spread out across the Americas and into the Atlantic and Pacific Oceans. The earliest year marked on any of these is 1848, which refers to the acquisition of California, New Mexico and Texas, at the end of the Mexican-American War (1846-48). Alaska was bought from Russia in 1867 (not 1876 as marked on the map) for $7.2 million, a purchase organised by the Secretary of State, William H. Seward – an act much mocked as 'Seward's Folly'. The 1890s saw the U.S. take over Cuba, Puerto Rico, the Philippines in the war against Spain, and separately Hawaii, all marked.

Further tentacles encircle areas of interest to the U.S. at the beginning of the war. Iceland was seen as vital for the protection of the Atlantic convoys and Britain had invaded it unopposed in 1940. While the island officially remained neutral during the war, its defence was transferred from Britain to the U.S. in 1941.[154] Equally essential to the war effort were the Dutch colonies of Suriname and Curacao. Over 85 per cent of the Allies' aviation fuel was refined here, as well as virtually all the fuel for the battles in North Africa.[155]

[154] Iceland's defence was handed over to the still neutral US by its British Commander-in-Chief, Maj-Gen Harry Curtis on 21 April 1941. Eventually over 40,000 American troops would be stationed here. Fairchild, Byron, 1960, *Decision to land United States Forces in Iceland, 1941*, Command Decisions, United States Army Center of Military History, p. 73.

[155] United States Government Printing Office, 1947, *Building the Navy's Bases in World War II: History of the Bureau of Yards and Docks and the Civil Engineer Corps, 1940-1946*, vol. 2, Washington, DC.

The text reads *"America lives by the law of the jungle. America for the Americans signifies that every country where the Yankees want to plant their flag, suddenly they are acting upon the Monroe Doctrine. The history of America is of imperialist plunder, where dollars were fired instead of bullets! It is the traditional policy of America, through the exploitation of European wars, to enrich themselves with the European possessions in the Atlantic and in the Pacific Ocean. Roosevelt also plans to use this war to benefit American world domination! [America] is seizing what the dying England has sacrificed!"*

In the Pacific Ocean the map stresses the might of the Japanese air force and navy, and in the Atlantic the German U-boats are systematically sinking American merchant ships, cutting the supply lines to Europe. The samurai sword severing the tentacle encircling the Philippines represents the successful invasion of the islands by the Japanese.

"But for the first time, America has miscalculated! Japan and Germany are encircling America in both oceans. The Pacific Fleet was destroyed at Pearl Harbor. In the Atlantic, warships and merchant ships are falling victim to the deadly underwater war. The tentacles of the Dollar Octopus will be cut. The gold of the international plutocracy, gathered in Fort Knox, is now surrounded by the irresistible armies of youth and the armies of labour."

In the upper-right corner is a paraphrase of a quote from *Life Magazine* by Henry R. Luce in the 17 February 1941 issue, writing about the American role in the world. Luce was the founder of *Time* in 1923 and *Fortune Magazine* in 1929, and was voted the most influential American of his day. He was a powerful supporter of Roosevelt in his desire to get the U.S. involved in the war effort.[156]

"In 1919 we had a golden opportunity, an opportunity unprecedented in all history, to assume the leadership of the world – an opportunity handed to us on the proverbial silver plate. We did not understand that opportunity. Wilson mishandled it. We rejected it. The opportunity persisted. We bungled it in the 1920s and in the confusion of the 1930s we killed it. To lead the world would never have been an easy task. To revive the hope of that lost opportunity makes the task now infinitely harder than it would have been before. Nevertheless, with the help of all of us, Roosevelt must succeed where Wilson failed."[157]

The artist, Louis Emile Manche, was a talented sculptor, ceramicist, graphic designer and painter, born in Amsterdam in 1908. He joined the Nationaal-Socialistische Beweging (the Dutch Nazi Party) in the 1930s and designed a number of posters for the NSB, whose propaganda department was directly controlled by Joseph Goebbels from Berlin. He also designed posters for the Dutch Waffen-SS, and indeed his brother Gerardus Martinus Manche, a Rottenführer in the SS, was killed near Kharkov on the eastern front in 1943.[158] Even for a committed Nazi such as Manche, the depths of treachery represented by this poster are startling. Not only produced at the behest of the occupying forces, it also extols the efforts of the Japanese forces that were currently overrunning Dutch colonies in the Far East and inflicting appalling cruelties on their inhabitants and their former masters. In the post-war period, Manche was interned with a number of other Dutch collaborators in the former SS concentration camp, Kamp Vught.[159] He died in Veldhoven in Holland in 1982.

[156] Herzstein, Robert Edwin, 2005, *Henry R. Luce, Time, and the American Crusade in Asia*, Cambridge, p. 1.

[157] Luce, Henry R., 17 February 1941, *Life Magazine*, New York, p. 64.

[158] www.oorlogsdodennijmegen.nl/persoon/manche/acf6f25c-2ccb-4633-bed6-1ff958e0ec0a

[159] Kamp Vught in Holland was with Natzweiler-Struthof in France, one of only two SS-controlled concentration camps in Western Europe sited outside Germany.

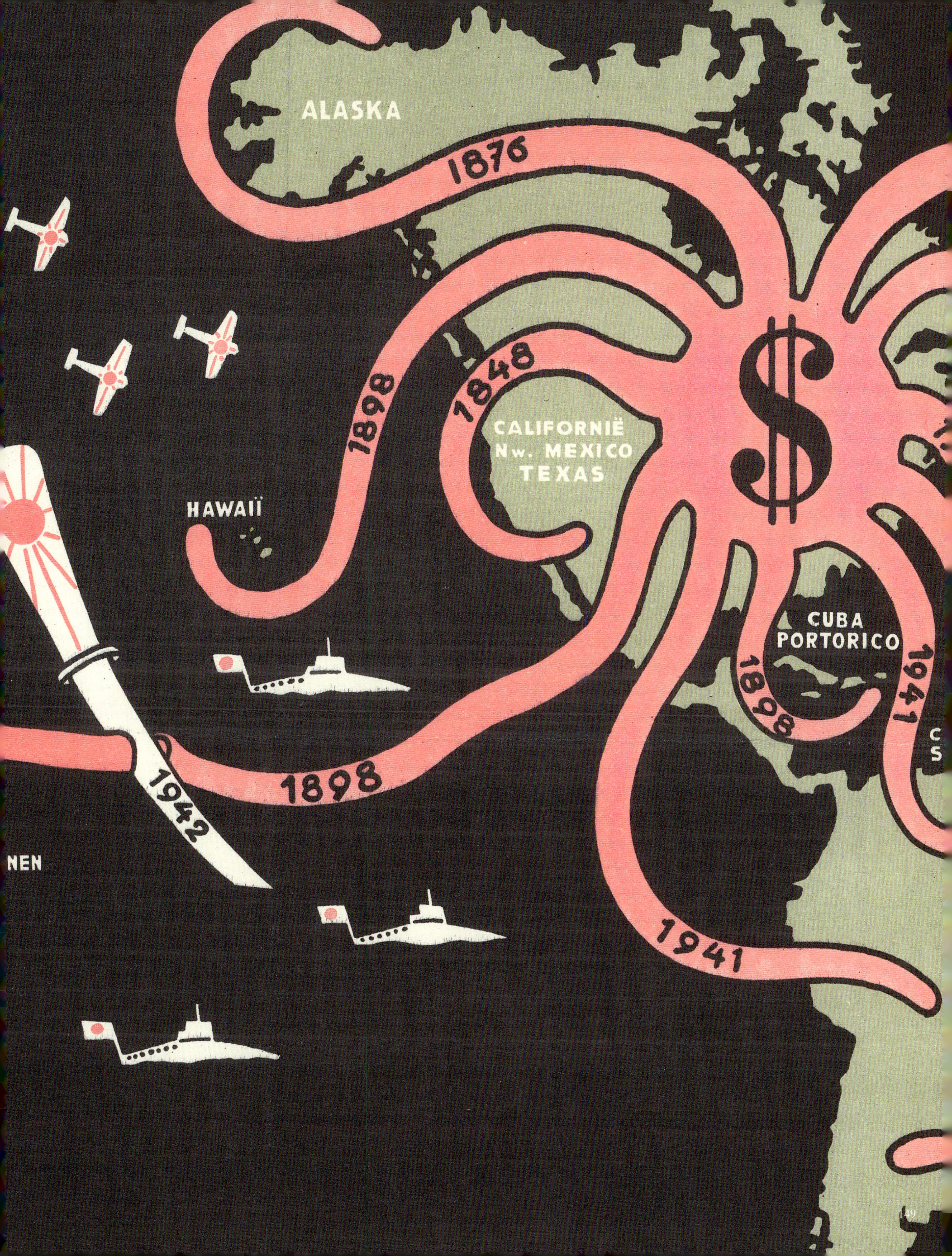
ALASKA
1876
1898
1848
CALIFORNIË
Nw. MEXICO
TEXAS
HAWAIÏ
CUBA
PORTORICO
1898
1941
1898
1942
1941
NEN
C
S

Anon: The War Map of the Great East Asia

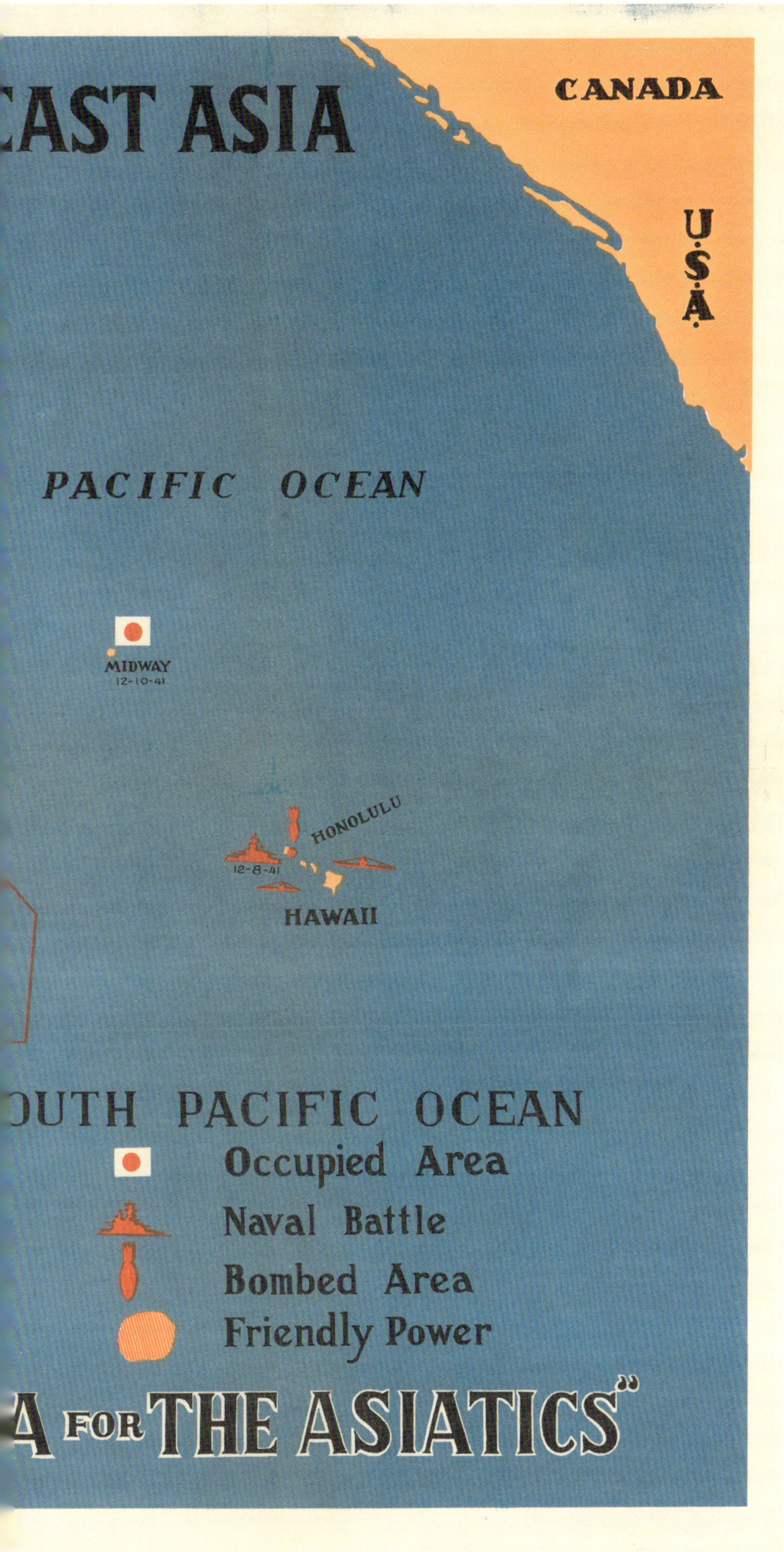

Anon: The War Map of the Great East Asia

lithographed map, [Japan], 22 February 1942,
35 x 20 in. (90 x 51 cm)

Published only a couple of months after the attack on Pearl Harbor, this is a simple but intriguing Japanese propaganda map showing the progress of the war in the Pacific.

The Japanese adoption of the term, "The Great East Asia War", and the use of slogans such as "Asia for the Asiatics" attempted to present the conflict in the Pacific as a struggle between the Imperial Forces trying to liberate their Asian brothers and sisters, and the brutal forces of colonial subjugation. It was part of Japan's efforts to garner the support or at least the grudging acquiescence from their subject nations for their conquest of the region.[160]

It is as a reflection on the hundreds of different languages in the areas already under the control of the imperial forces that Japan should have printed this piece of propaganda in English.

[160] Beevor, Antony, 2012, *The Second World War*, London, p. 277.

CALCUTTA
BURMA
PEGU
RANGOON
MORMEN
1-31-42
THAI
BANGKOK
FRENCH INDO CHINA
HANOI
HAINAN
SAIGON
HONG KONG
12-26-41
TAIWAN
APPARI
12-10-41
MANILA
1-2-42
ATIMONAN
PHILIPPINE
ISLANDS
GUA
BRITISH
MALAYA
1-11-42
KURLUMPUR
MEDAN
SHONAN
2-16-42
SUMATRA
KUTCHING
SARAWAK
12-16-41
PONTIANAK
1-29-42
BORNEO
2-10-42
PALEMBANG
2-15-42
BENKULENG
BALIKPAPANG
1-25-42
DAVAO
12-20-41
MENADO
1-11-42
BARJERMASIN
2-10-42
CELEBES
NETHERLANDS EAST INDIES
BATAVIA
2-4-42
MAKASSAR
1-27-42
CERAMI
1-31-42
SURABAYA
JAVA
BALI
2-22-42
DELI
2-21-42
KOEPANG
2-21-42
NEW
GUINE
DARWIN
AUSTRALIA

Hoping to capitalise on the deep and growing resentment caused by the European colonial exploitation of the peoples and resources of South East Asia, Japan proposed the creation of the 'Greater East Asia Co-Prosperity Sphere'. This claimed to be a planned confederation of South East Asian nations, united in harmony and equality, all striving for the common economic good, under the paternalistic leadership of Japan.[161] It was of course a fiction to disguise Japan's determination that it alone should be the power to replace the former colonial masters.

The situation shown in this map represents the high point in Japan's fortunes in the war. Korea, Manchuria and eastern China are in their hands. Hong Kong and Singapore have fallen. The Philippines and Borneo are occupied and are on the point of surrender. Sumatra and Java will fall shortly. The Japanese are everywhere triumphant, with all their principal enemies either defeated or in headlong retreat. Their forces are within reach of the very borders of British India and Northern Australia, and Darwin and Calcutta are already shown as having been bombed.[162]

Ironically the central feature of this map is the delineation of the Japanese Mandate. This was the area of the Pacific, principally the pre-First World War colonies of Germany, given over to Japanese control after the Treaty of Versailles as a reward for their adherence to the Anglo-Japanese Alliance and their declaration of war against Germany on 23 August 1914.[163]

Although in February 1942 the triumphant Japanese appear to be all-conquering, the tide of victory was about to turn irrevocably. This had been foreseen by Admiral Yamamoto, naval commander at the attack on Pearl Harbor, who declared before the attack that "I shall run wild considerably for the first six months or a year, but I have utterly no confidence for the *second and third years*." [164]

The first indication of what was to come was the Doolittle bombing raid by the Americans on Tokyo on 18 April 1942. Although the damage was minimal, the effect on the morale of the Japanese, who had long considered their homeland beyond the reach of their enemies, was shattering.[165] One unforeseen effect of the raid was the Japanese determination to push the U.S. carriers out beyond the reach of Tokyo that led to the disastrous decision by the Imperial Navy to invade the island of Midway. On 4-7 June the Japanese and American main carrier battle fleets met in the great pivotal battle of the Pacific War, the Battle of Midway. It was a decisive victory for the U.S. who sank all four Japanese carriers with the loss only of the previously damaged carrier, USS *Yorktown*. The Imperial Navy was unable to recover from the loss of half its entire carrier force and was forced to concede maritime control of the Pacific to the Americans.[166] The outcome could easily have been otherwise, as acknowledged by Nimitz in his report of the battle: 'Had we lacked early information of the Japanese movement, and had we been caught with Carrier Task Forces dispersed, possibly as far away as the Coral Sea, the Battle of Midway would have ended far differently." [167]

[161] Hoxson, Walter L. (ed.), 2002, American diplomacy in the Second World War, *The American Experience in World War II, vol. 8*, London, p. 42

[162] On 19 February 1942, 242 Japanese aircraft attacked Darwin in the first and still largest attack by a foreign power on Australia. See the National Archives of Australia, the Bombing of Darwin (www.naa.gov.au/collection/fact-sheets/fs195.aspx). Despite the claim on this map that Calcutta had been bombed, the city wasn't actually attacked until December 1942.

[163] Ponsonby-Fane, Richard, 1962, *Sovereign and Subject*, p. 348; The Class C mandate was given to former German possessions around the world, and were to be administered under the laws of the Mandatory as integral portions of its territory.

[164] Harry A. Gailey, 1995, *The War in the Pacific: From Pearl Harbor to Tokyo*, Novato, CA., p. 68.

[165] Sixteen B-25 *Mitchell* medium bombers were launched from the carrier, USS *Hornet* from far out in the Pacific. Too large to land back on their carrier, the planes of necessity had to fly on to Russia or China. Beevor, Antony, 2012, *The Second World War*, London, p. 307.

[166] Beevor, Antony, 2012, *The Second World War*, London, pp. 309-311.

[167] Commander-in-Chief Pacific Fleet to Commander-in-Chief Fleet, 28 June 1942, NHHC, *Battle of Midway*: 4-7 June 1942 Action Reports, *F-2042*.

Dudley Chase (Ernest): Total War Battle Map

lithographed map, Ernest Dudley Chase, Winchester, 1942, 36 x 16 in. (92 x 40 cm)

Dudley Chase (Ernest): Total War Battle Map

For the first time in history, the years 1939 to 1945 saw an all-encompassing World War involving every inhabited continent. This American pictorial map by Ernest Dudley Chase shows the world in 1942, a critical point in the fortunes of the Allies.

The Axis powers of Germany and Japan, including the Japanese occupied territories of Korea and Taiwan, are in ominous black. The omission of Italy is surprising, since it did not withdraw as an Axis Power until the Allied invasion in September 1943. Stalingrad, one of the most critical battles of the war and still raging when this map was drawn, is singled out in red. The borders are pre-war, the possessions of the main powers are shown by symbols referenced in the index in the lower left, and important ports and railways are also marked.

The oceans are filled with detailed vignettes of naval ships and aircraft, including a US Naval K-class blimp floating above Bermuda. There is a note above the Mindanoa Deep: *"Greatest Known Ocean Depth, 35,400ft Depth"* that *"This would be a good place to sink the Jap fleet"*.

There are also six insets showing a selection of weapons the Allies had available in the air and on land: coastal artillery, howitzers, anti-aircraft guns, a B-17 Flying Fortress and a 'Yank Tank', the M3 Lee.

Dudley Chase added a small comment to the title regarding the design: "A battle map, to set the stage for 'Total War' on land and sea; and, in the border, just a hint – a prayer – for total victory!" This hint refers to the 'V' for Victory signs placed in each corner of the map, and V (...–) repeated around the border in Morse code.

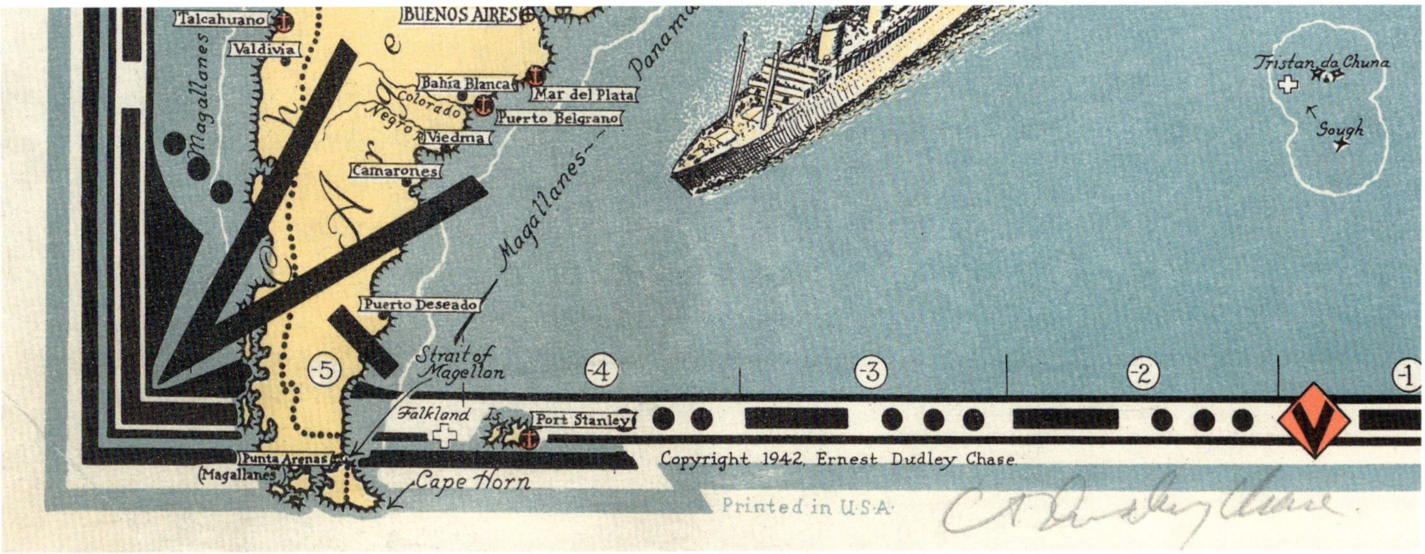

Ernest Dudley Chase (1878-1966), born in Lowell, Massachusetts, is one of the best known and prolific pictorial mapmakers of the twentieth-century, creating more than 50 different maps in his very distinctive style.[168] A perfectionist, he was noted for his fine detail and vignettes with explanatory captions.[169] They frequently dealt humorously with serious subjects while always reflecting his passionate patriotism. Many of his maps depict places in his native north-eastern USA, but he also designed maps of the world, North and South America, various countries in Europe, and numerous war maps.[170] A number of his maps, including this one, were signed by Chase in pencil. In 1945 Tim Thrift wrote a short biography of Dudley Chase, in which he discusses the importance of his artistic output: *"Thousands of his 'Total War' maps have been used to follow the course of our armies and navies abroad. In fact, it is anticipated that many of the Chase maps – such as those for the British Isles, Italy, France, Germany, etc. – will be in great demand after the war by our men in the armed forces to show just where they have been, as the illustrations will mean much more than just pictures to them – they were in many of the buildings shown, saw the* 'points of interest' *with their own eyes!"*[171]

Dudley Chase was educated as a graphic artist and began his career by designing greeting cards. He sold his own company to Rust Craft Publishers in 1920 and he subsequently held various managerial posts there until his retirement in 1958. He wrote the book: *The Romance of Greeting Cards* in 1926, and started drawing maps in the early 1930s.[172] His third wife, Clara Katrina Holland Chase, was a celebrated pictorial mapmaker in her own right.

[168] http://mastatelibrary.blogspot.co.uk/2009/05/ernest-dudley-chase-worldview-in-maps.html

[169] Thrift, Tim, 1945, *A Meticulous Maker of Maps*, Boston.

[170] The Harvard Map Collection, to whom Ernest Dudley Chase donated examples of most of his maps, presented an exhibition of these in March 2003, curated by Joseph Garver. Later, from May to September 2009, the State Library of Massachusetts presented the exhibition: "*Ernest Dudley Chase: A Worldview in Maps*".

[171] Thrift, Tim, 1945, *A Meticulous Maker of Maps*, Boston, pp. 9-10.

[172] http://mastatelibrary.blogspot.co.uk/2009/05/ernest-dudley-chase-worldview-in-maps.html

Anon: Mappa de Europa

Anon: Mappa de Europa [Map of Europe]

lithographed map, Lisbon, c.1942, 18 x 14in (46 x 36cm)

This unusual and very scarce Portuguese pictorial depicts the Second World War as a straightforward conflict between mainland Europe and Soviet Russia.

The neutrality of Portugal under the rule of its dictator, António de Oliveira Salazar, was a more complex balancing act than it might at first appear. Salazar, while by nature conservative and authoritarian, loathed what he regarded as the barbarism of Nazi Germany. There had also been an almost unbroken alliance between Britain and Portugal for over six centuries. While Salazar offered to come to Britain's aid if asked, both he and Churchill were acutely aware that if Portugal declared for the Allies, Spain would immediately join the Axis Powers with the consequent likely loss to the Allies of the whole Iberian Peninsula.[173]

The message of this propaganda piece is overtly pro-Axis and anti-communist. A German soldier is seen attacking the monstrous figure of Soviet Russia which is destroying churches and killing local populations. Numerous other figures represent European countries coming to Germany's aid presumably as members of volunteer brigades of the SS as many did, including such unlikely supporters as Denmark, Belgium and Holland.

The only figure ignoring the struggle is that of Portugal itself who stares pointedly out to sea. Britain, a major supplier of aid to Russia, is in a close embrace with a Soviet soldier shown while surrounded by sinking merchant ships.

Despite the message of this map, Salazar never wavered in his belief that Britain would remain unbeaten and that with American help, Germany would ultimately be defeated. Portugal would use her position of neutrality to aid the thousands of refugees that poured across her borders.

[173] Leitz, Christian, 2001, *Nazi Germany and Neutral Europe during the Second World War*, Manchester, p. 149.

The "TIME & TIDE" Map of TH

COPYRIGHT, 1942

PRINTED IN GR

E ATLANTIC CHARTER

PHILIP & SON, LTD., LONDON

GEORGE PHILIP & SON, LTD., THE LONDON GEOGRAPHICAL INSTITUTE

Gill (Macdonald): Time & Tide - Atlantic Charter Map of the World

They sha
into pl
spears
nation
sword
neithe
wa

Gill (Macdonald): Time & Tide – Atlantic Charter Map of the World

lithographed map, George Philip & Son, Ltd., London for Time & Tide Magazine, 1942, 43 x 34 in. (109 x 87 cm)

This magnificent map drawn up while the Second World War still raged is already envisaging the post-war world following an Allied victory.

Beneath the large scroll listing its aims, the Atlantic Charter is shown radiantly illuminating the world and dispersing the clouds of conflict.

The focus is on natural resources and the web of shipping lanes, signifying that a global trade network could create a peaceful world.

In the lower left-hand corner, the clouds have rolled back to reveal a powerful figure destroying the weapons of war with an idyllic farming scene behind illustrating Isaiah (2:3-4) 'They shall beat their swords into ploughshares and their spears into pruning hooks: nation shall not lift up sword against nation, neither shall they learn war anymore."

The Atlantic Charter was a statement of general principles to govern the post-war world agreed between Winston Churchill and Franklin D. Roosevelt on the USS *Augusta* in Placentia Bay near Newfoundland on 9 August 1941, and issued as a joint declaration one week later.[174] It became the basis for the United Nations with the aim to define the goals for the post-war period.[175] All the nations making up the Allies of the Second World War agreed on the charter's principles and signed the Declaration of the United Nations on 1 January 1942, shortly after the attack at Pearl Harbor and the entry of the United States into the war.[176] The charter was inspired by Woodrow Wilson's ideas when setting up the League of Nations 25 years earlier, and the map pays homage to these ideals and the belief in a lasting peace. It is in fact only paragraph six that refers directly to the ongoing war, by mentioning "the final destruction of Nazi tyranny".

Macdonald Gill was a multi-talented artist. He initially trained as an architect but also mastered painting, cartography and type and poster design. Inspired by the Arts and Crafts movement, he was considered a leading figure among the 'Medieval Modernists'.[177] His first professional map was finished in 1909, where he had already established his unique style combining simplicity of colour with masterfully controlled complexity of design. His most influential map was the grand pictorial map he did for London Transport known as the *Wonderground Map*, which inspired numerous copies and influenced a generation of pictorial mapmakers throughout the world.[178]

In 1917 Max Gill was invited onto the Headstone Design Committee of the Imperial War Graves Commission along with his brother Eric, and created the font seen on all the British and Colonial gravestones from both world wars.

[174] The statement issued with the declaration can be seen on the map within the circular ribbon attached to the charter, centred on the right-hand side.

[175] Beevor, Antony, 2012, *The Second World War*, London, p. 220.

[176] Whitfield, Peter, 2010, *The Image of the World*, London, p. 132.

[177] Personalities in the Realm of Art, Macdonald Gill, F.R.I.B.A, *Town and Country News*, 29 September 1933.

[178] Burdon, Elisabeth, 2009, Macdonald Gill: *The "Wonderground Map" of 1913 and its influence*, *IMCoS Journal*, vol. 116, p. 9.

Atlantik-Charta

Anon: Atlantik-Charta

Anon: Atlantik-Charta [Atlantic Charter]

silver bromide print, Germany, c.1942, 23 x 13in. (58 x 33cm)

This is a rare example of a propagandist responding directly to the message and image of an opposing piece of propaganda published by the enemy.

The anonymous German artist has clearly had access to Macdonald Gill's *Time and Tide Map of the Atlantic Charter.* Furthermore he has even copied the facsimiles of Roosevelt's and Churchill's signatures that Gill had incorporated into the earlier map.

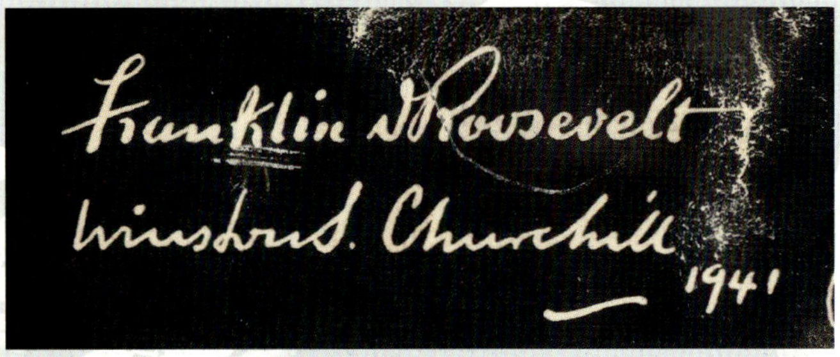

This map also lists the eight principal points in the charter, here combining each of the points with text and images ridiculing the lofty sentiments behind them. The tone is set by the title cartouche purporting to be Roosevelt and Churchill's declaration that *"The Atlantic Charter is our Holy Will. We are the only ones who are able to implement it. This is proved by the many wonderful achievements we can show from our own countries. Anyone laughing at this is unworthy of our control."*

Significantly mainland Europe is missing, and has been replaced by text saying: *"Undiscovered territory (soon to be sold to Mr. J. Stalin)."* In the South Pole, the penguins are commenting on point 2 of the charter: *"They [the Allied Powers] desire to see no territorial changes that do not accord with the freely expressed wishes of the peoples concerned"*, by stating: *"We are the only ones that still believe it."*

The British Empire was a frequent subject for German propaganda often contrasting the size of Britain's overseas possessions with her "hypocritical" attempts to limit the territorial expansion of others. A British colonial administrator in India proclaims: *"No one does colonial administration quite like us Englishmen."* A further note describes the Australian as being *"at war only in theory"*, while in Africa the British rulers are leading the native population off in chains to fight the war for them.

A repeated image on this map is that of Uncle Sam, or other figures representing the US, using their wealth and power to bully and intimidate other nations including their ally, Britain.

It is not surprising that the eight points of the Atlantic Charter should so enrage the Nazi authorities. Their avowed purpose was to form the basis of restructuring the world order following an Axis defeat. What is much more unusual is to have the opportunity through this and the Gill map, to be able to look at both opposing views of a single subject.

The usual elements of mapmaking, the wind-rose and the scale-bar have also taken new forms in this map. The wind-rose features Stalin's head blowing hammer and sickle symbols in all directions. The scale bar, seen as a rifle in Lehmann-Dumont's maps of Europe in 1914, is here a broken stick representing the very much reduced strength of the Allies.

Blake (Frederick Donald): A Batalha do Atlantico

Blake (Frederick Donald): A Batalha do Atlantico [Battle of the Atlantic]

Alf Cooke Ltd., Leeds and London, 1943, 21 x 31in. (53 x 79cm)

This British propaganda map by the artist, Frederick Donald Blake, brilliantly evokes the situation at the crucial moment the balance of power in the Atlantic swung irresistibly in favour of the Allies.

The Battle of the Atlantic was the longest continuous military campaign of the Second World War and has been called the "longest, largest, and most complex" naval battle in history.[179]

At the beginning of the war, the U-boat blockade was seen by both sides as having a very real possibility of cutting Britain off from the essential supplies and food necessary to conduct the war and forcing her into capitulation. Churchill would later famously write: *"...the only thing that ever frightened me during the war was the U-boat peril."*[180]

At the beginning of the battle, the advantage seemed to lie with the U-boats; indeed the period of June 1940 to February 1941 was called 'Die Glückliche Zeit' (The Happy Time) by their exultant German crews.[181] The principal problem for the submariners was finding the convoys in the vastness of the ocean. This was at first partly overcome with the use of long-range spotter planes based in Norway, who then directed the wolf packs onto their virtually unprotected targets using Enigma coded radio messages. During the Happy Time, the British and their allied merchant fleets were to lose almost 300 ships, amounting to nearly 1,500,000 tons.

From 1941 onwards the Allies began to redress the balance in their favour. Firstly and most importantly was the introduction of the convoy system. This followed the realisation that one very large convoy, even unescorted, was much harder to find than numerous small ones. The next crucial factor was the breaking of the Enigma codes by the cryptographers of Bletchley Park, which allowed the Admiralty plotters to steer the convoys away from where the U-boats were concentrating.[182]

After these principally defensive measures, the Admiralty began take the battle to the U-boats themselves. One significant development for the Royal Navy was the introduction of the 'Hedgehog' depth charge. This addressed the chief weakness of the ASDIC (or Sonar submarine detection system) which would lose contact with its target at close range, as the sound of the attacking ship's own engines would mask the position of the U-boat below just when it was within range of dropped depth charges. A cunning U-boat commander would change course the moment the ship was overhead. The forward-firing 'Hedgehogs' though could be launched well ahead of the destroyer while the position of the U-boat was still being tracked.[183]

All along the Atlantic Wall, the Allies are shown attacking the German supply lines in 'Fortress Europe' and the important U-boat pens. Hamburg, Wilhelmshaven, Bremen, Brest, Lorient and St. Nazaire are all ablaze. *"The U-boats' Atlantic Coast bases are being pounded by the Allied Air Forces and the entrances to their harbours are being mined from the air. The factories where they are being built are being crippled by bombs."* [184]

179 Syrett, David, 1994, *The defeat of the German U-boats: The Battle of the Atlantic*, Columbia, SC.

180 Costello, John & Hughes, Terry, 1977, *The Battle of the Atlantic*, London; Beevor, Antony, 2012, *The Second World War*, London, p. 182.

181 The dates vary according to different sources.

182 Beevor, Antony, 2012, *The Second World War*, London, pp. 183-184. The German Kriegsmarine added an extra rotor to the naval Enigma encoder and throughout the war considered its codes unbreakable.

183 www.uboat.net/allies/technical/asdic.htm

184 Translation of part of the text in the panel in the lower right-hand corner with the headline: "*A Ceaseless Battle is Raging in the Atlantic.*"

Another crucial factor in the reduction of the U-boat menace was the closing of the mid-Atlantic 'Greenland Gap', the area that land-based anti-submarine aircraft had previously been unable to reach. By 1943, the introduction of Long Range Liberator bombers flying from Newfoundland and Iceland effectively closed the gap, and no longer gave the U-boats the respite of safe surface sailing.[185]

The Allied air supremacy by early 1943 is represented by the waves of aircraft surrounding the British Isles. Stricken U-boats of the Kriegsmarine are interspersed between the undamaged lines of Allied supply ships. May 1943 became known as 'Black May' to the submariners and the catastrophic loss of 43 U-boats in this month alone prompted Dönitz to call off all operations in the North Atlantic on 24 May. The Battle of the Atlantic was over and Churchill's greatest worry was behind him.[186]

Designed to encourage supporters of the Allies as well as wavering neutrals, this map was published in English, French, Dutch, Arabic and as here, Portuguese.[187]

Frederick Donald Blake was born in Scotland in 1908, but the family soon moved to London where he attended the Camberwell School of Art. He eventually found work as an architectural draughtsman, but also started to experiment with watercolours and painting in oils. Blake, like a number of his contemporary artists was drafted by the Ministry of Information during the Second World War to produce propaganda material including maps and pictorial pieces. After the war he continued painting and was in great demand as a commercial artist, working for the railways, bus companies and the aircraft industry. He died in 1997.[188]

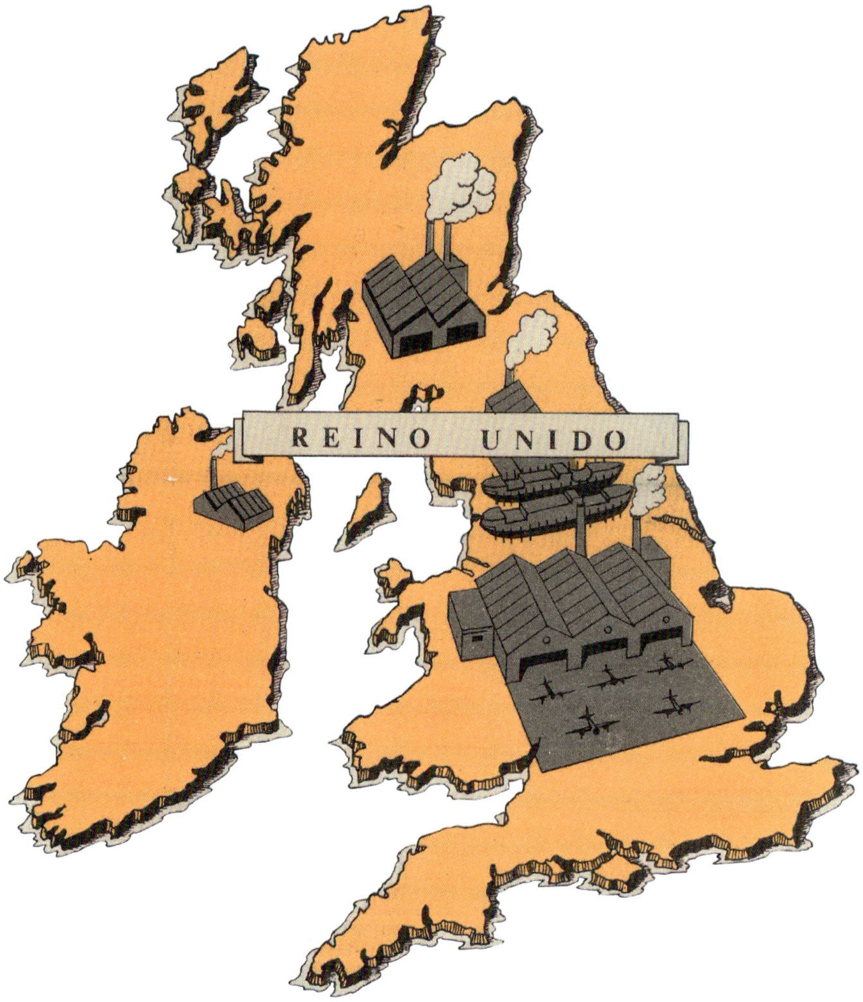

[185] Despite its name, the U-boat was principally a surface vessel, only submerging when attacked or attacking. It was particularly vulnerable to aerial attack.

[186] Beevor, Antony, 2012, *The Second World War*, London, p. 438.

[187] Bryars, Tim and Harper, Tom, 2014, *A History of the 20th Century in 100 Maps*, London, p. 103.

[188] www.fdonaldblake.com/

COMBÓIOS VINDOS DO
CANADÁ e dos
ESTADOS UNIDOS
COMBÓIOS E AVIÕES VINDOS DO
CANADÁ e dos
ESTADOS UNIDOS

The Star Weekly, Toronto, August 1, 1942 19

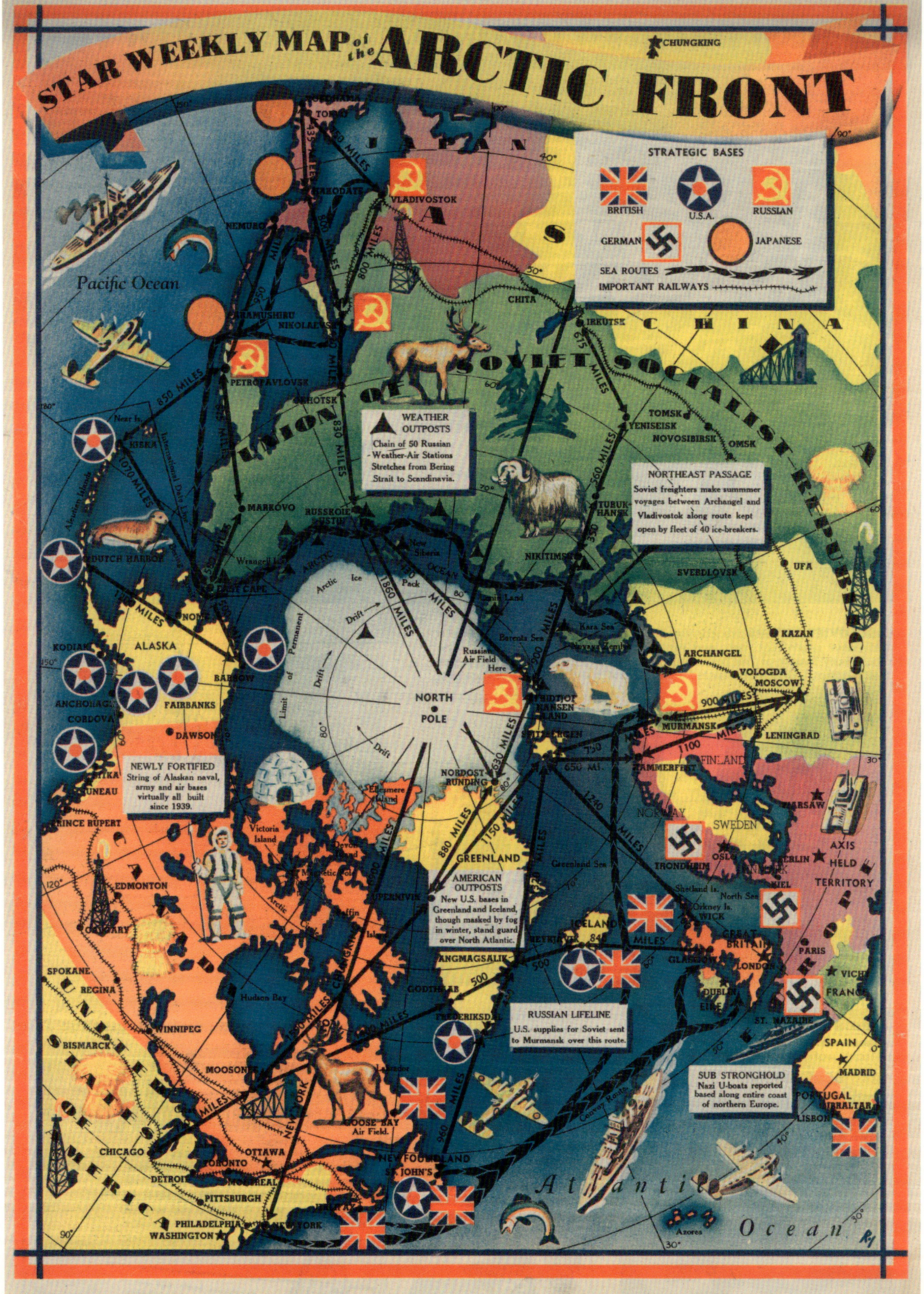

THOSE ACCUSTOMED to maps of the eastern and western hemi spheres will be surprised to find the earth's great land masses as close to each other in the north as indicated by this Star Weekly map of the Arctic front. Here is a bird's-eye view of the earth from directly over the north pole, showing the permanent ice pack of the Arctic, the shipping lanes which skirt it, and the air lanes, some already used, which cross it. This map reveals that the shortest route to Tokyo from Vancouver is northward along the British Columbian coast; also the shortest air lines between North American centres and Asiatic cities are across the "frozen roof of the world." The Arctic will probably give up many of its secrets to postwar aviation developments.

R.I.: The Arctic Front

chromolithographed news print map, The Star Weekly, Toronto, 1 August 1942, 10 x 15 in. (25 x 38 cm)

Like Richard Harrison's map of the previous year, again a polar projection is used to highlight the proximity of the principal combatants to each other.

The situation had changed dramatically since the publication of Harrison's map of 1941. The U.S., no longer neutral, was now engaged in a global conflict and, through lend-lease, was the major supplier of the means of war to her allies. The need to supply both Britain and Russia in the U-boat infested northern waters, together with the greatly increased range of aircraft, had made the great wastes of the Arctic landmass a significant factor in warfare for the first time in history.

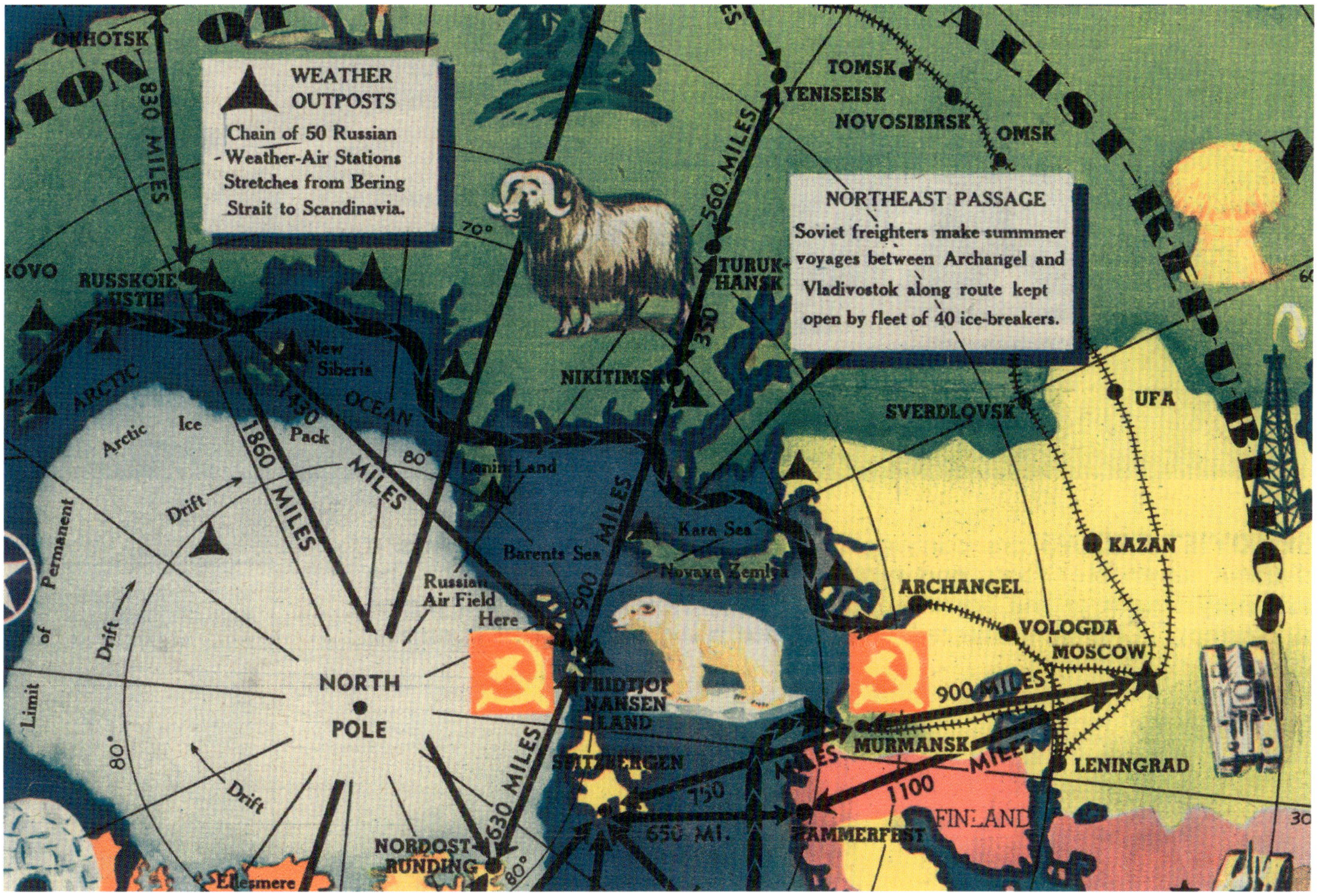

This map also touches on a strange and now almost forgotten campaign fought within the pitiless confines of the Arctic Circle. This was a long-running battle by the Russians, Germans and Norwegians to establish meteorological bases on Svalbard. Weather data from the Arctic was vital for both the Allies and Axis powers for predicting weather patterns in Central Europe and the Northern Atlantic. The Svalbard peninsula was particularly important because of its proximity to Murmansk, the destination of the British and American convoys bringing essential supplies to Russia. In 1944-45, despite the deteriorating position for the Germans in Europe, they were still maintaining four manned weather stations in Svalbard. Indeed the very last German unit of the Second World War active anywhere in the world was manning the weather station on Nordaustland on Svalbard. They were eventually evacuated to their great relief by the Norwegian navy in September 1945, nearly four months after the end of the war in Europe.

Kimon Evan Marengo (Kem): Une a Une, on lui Brisera les Pattes

Kimon Evan Marengo (Kem): Une a Une, on lui Brisera les Pattes [One by One, His Legs will be Broken]

lithographed map, London, c.1943, 20 x 30 in. (50 x 76 cm)

This vivid British propaganda map was directed at the various peoples in the previously French controlled colonies of North Africa. It can be seen as the counterpart to the *Confiance – ses amputations se poursuivent méthodiquement* [see map on p.136]. Hitler, a spider, squats malignantly at the centre of Europe. His legs reach out to North Africa, the Soviet Union, Scandinavia and the Atlantic Ocean, where they are being attacked and severed by Allied aircraft, tanks and warships.

In contrast to the powerful and threatening octopus tentacles of the earlier propaganda pieces, KEM subtly emphasizes the fragility of the Nazi's grasp with his use of the spider. The strong image and clear simple message, without lettering other than the title, allows for its use across the different languages of Europe and Northern Africa.

Kimon Evan Marengo was a British cartoonist, born in Egypt, the son of a Greek cotton merchant. After attending the Ecole des Science Politique in Paris, he went to Oxford University but his studies were interrupted by the outbreak of war. His graphic ability and wide knowledge of foreign languages made him a natural for wartime propaganda and he joined the Ministry of Information, became involved with *Political Warfare Executive* and eventually rising to the rank of Lt. Colonel.[189]

[189] The highly clandestine Political Warfare Executive was Britain's principal organisation for the conduct of psychological warfare against Nazi Germany and its allies. The philosophy of the PWE was perhaps best expressed by Sefton Delmer, one of its most energetic practitioners, who declared: *"We must never lie by accident, or through slovenliness, only deliberately"*. The National Archives, 2005, *Allied Propaganda in World War II: The Complete Record of the Political Executive*, London.

Blake (Frederick Donald): Britain - Spearhead of Attack

Blake (Frederick Donald): Britain - Spearhead of Attack

lithographed map, Alf Cooke Ltd., Leeds and London, c.1944,
29 x 19 in. (74 x 48 cm)

This is another British propaganda map by the artist Frederick Donald Blake, again celebrating Allied successes in Europe.

"Night and day the bombers of the R.A.F. strike deep into the enemy's war machine… Almost daily R.A.F. fighters sweep over enemy occupied territory to strafe communications and strategic targets…"

Identical in style to Blake's earlier map of the Battle of the Atlantic, but now Allied attacks are reaching even further into mainland Europe. Britain is shown in a brighter colour than the rest of the map highlighting its resources and production facilities: food, coal and electric power, as well as aircraft, ships, tanks and guns. Strategically important continental targets such as bridges, railways, power stations, gas works, general factories and troops are generically indicated, while some industrial complexes such as the Renault Works are individually shown. Bomber Command's reach and power was extended by the introduction of four-engined Halifaxes and Lancasters, and on 3 March 1942, 235 bombers attacked the Renault Factory at Boulogne-Billancourt on the edge of Paris in the first mass attack on a target in France. [190]

There is an interesting inset of Northern Italy showing the bombing of Genoa, Turin and Milan. Unlike in France where only military and industrial objectives were targeted, the Allies deliberately bombed Italian civilian centres to break morale and encourage disillusionment with Mussolini's regime.[191]

In the North Atlantic, several landing craft are shown heading for Lofoten and Vaagsö, commemorating Operations Claymore and Archery, two of the earliest British Commando raids on the Norwegian Coast. [192]

[190] Beevor, Antony, 2012, *The Second World War*, London, p. 443.

[191] See: Bombing, States and Peoples in Western Europe 1940-1945, *The Bombing of Italy 1940-45* (http://centres.exeter.ac.uk/wss/bombing/italy.htm)

[192] Operation Claymore: In March 1941, a major commando raiding party landed at Lofoten and destroyed 18,000 tons of shipping and 3,600 tons of fish oil and glycerine used by the German war industry. Other than the substantial and timely propaganda value, the most significant result was the capture of an Enigma rotor arm allowing the decrypters of Bletchley Park to break the German naval codes. See: Chappell, Mike, 1996, *Army Commandos 1940–1945*, London.

The cities of Central Germany are depicted as being uniformly ablaze. Several of the largest bombing raids, such as those on the Philips Radio Works and the Schneider-Crusot Factory, are marked individually. According to the map, the furthest east the bombers of the Royal Air Force reached was Pilsen in the Czech Republic and Danzig in Poland.[193]

There was a passionate debate at the time as to the morality and effectiveness of the Allied bombing strategy against German cities which, of course, continues to this day. The principal proponent of the area bombing of civilian centres as a means of destroying morale was Arthur 'Bomber' Harris, head of the RAF's Bomber Command. The U.S. Army Air Force, under the command of Brigadier General Ira C. Eaker, advocated the precision bombing of 'key nodes' of the enemy's 'industrial fabric'.[194] Unfortunately so-called precision bombing in daylight proved to be extremely costly to the attacking aircraft, and night-time bombing was almost wholly ineffective in terms of accuracy. The secret Butt Report commissioned by the British in August 1941 found that only one in five aircraft was succeeding in dropping its bombs within five miles of the target. The conclusion drawn was that only targets the size of cities stood any realistic chance of being hit.[195]

In retaliation for the bombing of German cities, Hitler ordered the bombing of a number of historic cities in England including Bath, Canterbury, Exeter, Norwich and York. This became known as the Baedeker Blitz after Baron Gustav Braun von Stumm, a spokesman for the German Foreign Office, is reported to have said on 24 April 1942 (following the first attack): "We shall go out and bomb every building in Britain marked with three stars in the Baedeker Guide." [196]

A further aspect of the war at sea is featured in the images of mines being parachuted into the North and Baltic Seas. In twenty months after the introduction of aerial mines dropped by parachute in April 1940, they accounted for 164 Axis ships with the loss of 94 Allied planes. Direct aerial attacks on Axis shipping during the same period accounted for 105 ships with the loss of 373 planes. During the course of the war, the UK dropped a total of 48,000 mines into the sea in the European theatre alone.[197]

[193] The Pilsen raid targeting the massive Skoda Armaments plant was ordered by Harris in May 1943. It was at the very edge of the operational range of the bombers. Tragically the 285 bombers missed their target by some twelve miles and, mistakenly identifying a large mental hospital for the Skoda Works, deluged the area with 691 tons of bombs (www.globalsecurity.org/military/world/europe/skoda-pilsen.htm)

[194] Beevor, Antony, 2012, *The Second World War*, London, pp. 445-447.

[195] See: Longmate, Norman, 1983, *The Bombers: RAF offensive against Germany 1939-1945*, Hutchinson, London.

[196] Price, Alfred, 1977, *Blitz on Britain*, Abingdon, p. 132.

[197] Youngblood, Norman, 2006, *The Development of Mine Warfare: A Most Murderous and Barbarous Conduct*, Greenwood, pp. 129-130.

KIE
EMDEN
BREMEN
HAMB
OSNABRÜCK
MÜNSTER
HAMM
GELSENKIRCHEN
DORTMU
ESSEN
DUISBURG
KREFELD
DÜSSELDORF
COLOGNE
Philips Radio Wks.
Airfields
LOGNE
Power Stn.
Factories
Troops
Bridges
Renault Wks
Matford Wks.
Schneider-Creusot Wks.
LUDWIGSHAFEN
MANNHE
SAARBRÜCKEN
KARLSRUHE

Johnson (Major Tom Loftin): Map of West Point

Johnson (Major Tom Loftin): Map of West Point

lithographed map, West Point, New York, 1944,
28 x 22 in. (71 x 56 cm)

This American pictorial celebration of the history of West Point consciously links the current American wartime leaders, Eisenhower and MacArthur with the great military figures of the past.

The United States Military Academy at West Point is arguably the most famous military academy in the world. George Washington considered the site of the current academy, situated on a plateau on the Hudson River 50 miles (80 km) north of New York City, the most important strategic position in America. Washington had Thaddeus Kosciuszko construct a fortification in 1778 and transferred his headquarters there the following year. It was never conquered by the British, and is therefore the oldest continuously occupied military post in America.[198]

It was Thomas Jefferson who signed the legislation that established it as the United States Military Academy in 1802. Officers educated there served with distinction in both the Mexican and Indian Wars, and the most famous commanders on both sides in the Civil War were graduates from West Point.

This map shows the expanse of the Academy in 1944, from the Lee and Washington Gates in the north to the Thayer Gate in the south. In the centre are The Plain Parade and the flag pole, while the majority of the roads and buildings are named after American heroes. Scattered over the map are notes on events and characters that indirectly helped shape the Academy. One tells us that West Point was purchased from Stephen Moore in 1790, while another mentions the death of Benny Havens in 1897.[199]

[198] The official West Point website: www.usma.edu/wphistory/SitePages/Home.aspx

[199] In fact he died in 1877.

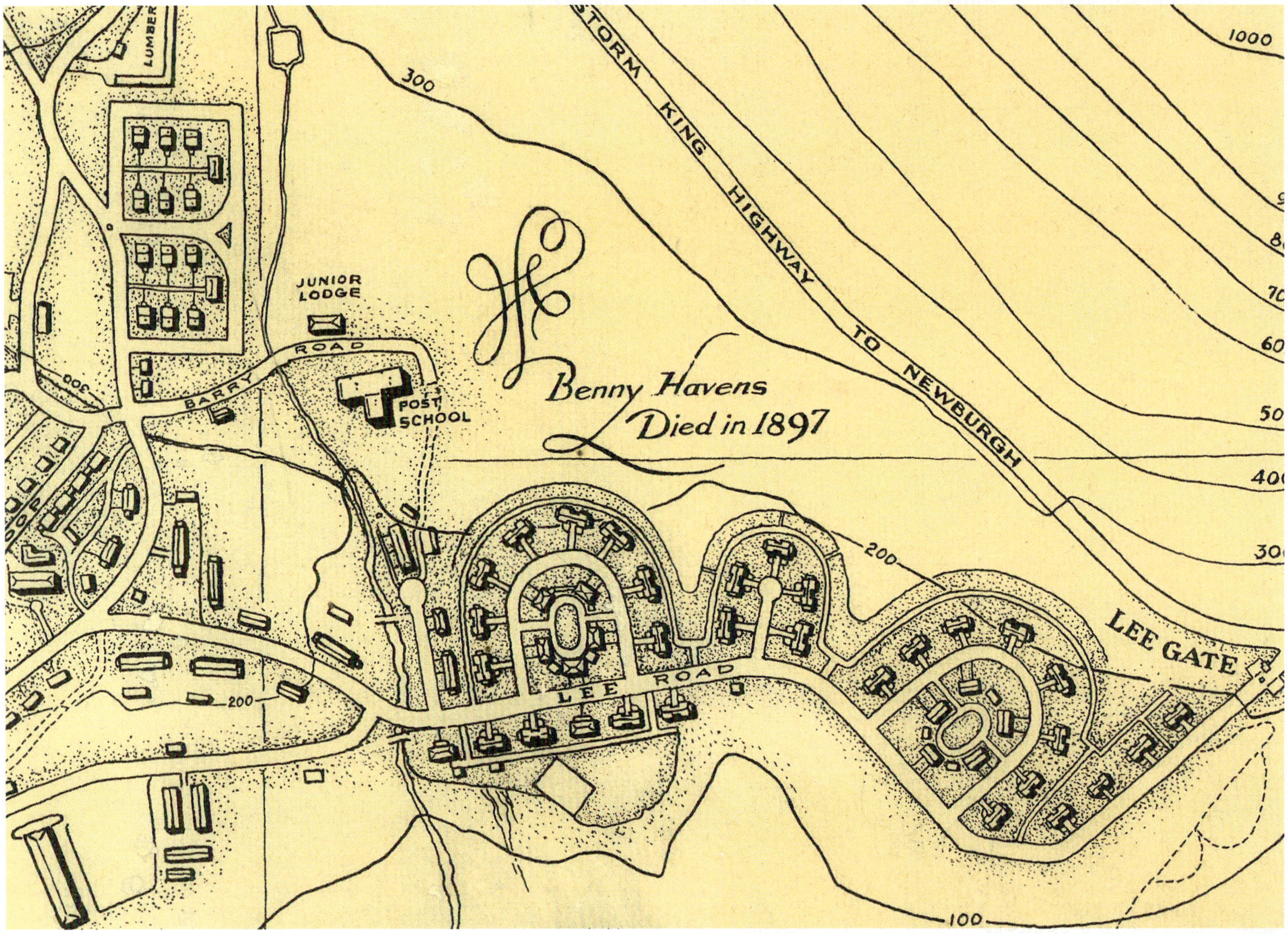

Havens was a legendary figure to generations of West Point cadets and a song commemorating him is still sung in the Academy today, *Benny Havens, Oh!*.[200] Drinking in the Academy was illegal, but there were still a number of taverns in its vicinity in the 1820s and 1830s, of which the most popular were Gridley's and Benny Havens's located in Highland Falls. Havens set up his tavern in 1824, and ran it with his wife, Letitia for over thirty years. Edgar Allan Poe, who spent a brief and deeply unhappy period as a West Point cadet, commented that Benny Havens "was the sole congenial soul in the entire God-forsaken place."[201] The Havens were also the only two American citizens forbidden to set foot on the land of West Point, after Sylvanus Thayer, 'the Father of West Point', caught them supplying liquor to the cadets. Until 1825 the two taverns were in competition, but Thayer persuaded Congress to buy Gridley's in order to close it down, and so inadvertently gave Havens a monopoly.

Bordering the map are the names of some of the most famous graduates to come out of West Point: Sylvanus Thayer, Robert E. Lee, Grant, Sherman, Custer, Stonewall Jackson, Goethals, Eisenhower, MacArthur, Pershing, Meade, Pickett, McClellan, Schofield and Arnold.[202]

[200] www.west-point.org/greimanj/west_point/songs/bennyhavens.htm

[201] Wood, Robert J., 1937, Early Days of Benny Havens, *The Pointer*, 26 February. Sadly the often told story that Poe, when ordered to parade armed and wearing cross-belts, appeared carrying his musket and wearing only the belts, is entirely apocryphal. He was in fact expelled for disobedience and gross neglect of duty having managed to rack up a record of 60 demerits in under a month.

[202] Thanks to Michael Buehler for pointing out that the Arnold referred to was Henry 'Hap' Arnold, the only American to hold the rank of five star general in two different services (General of the Army and General of the Air Force) rather than Benedict Arnold, the notorious traitor, who was not a graduate but did attempt to sell West Point to the British for £20,000.

George Armstrong Custer was notorious for his less-than-glorious tenure at West Point. He graduated bottom of his class of thirty-four and notoriously amassed a record number of 726 demerits.[203] Despite this, after his body was recovered from its makeshift grave at the site of the Battle of the Little Bighorn, it was solemnly reinterred in the cemetery at West Point.

The artist, Major Tom Loftin Johnson, was the art instructor at West Point. In 1936 he created an enormous mural entitled *Panorama of Military History* for the cadets' Gothic mess hall. Measuring 35 x 71 feet and produced using the ancient technique of egg tempura on plaster, it is one of the largest unbroken interior pictures in the world and took Johnson over a year to complete.[204] So enamoured was he with West Point that he asked that his cremated remains should be placed in the wall below the mural with a simple plaque declaring, *"He gave his best to West Point"*. Sadly for Johnson this request was never honoured.

[203] Eicher, John H. & Eicher, David J., 2001, *Civil War High Commands*, Stanford, p. 196.

[204] *Time Magazine*, 20 July 1936.

U.S. Navy, Bureau of Naval Personnel: The Mediterranean

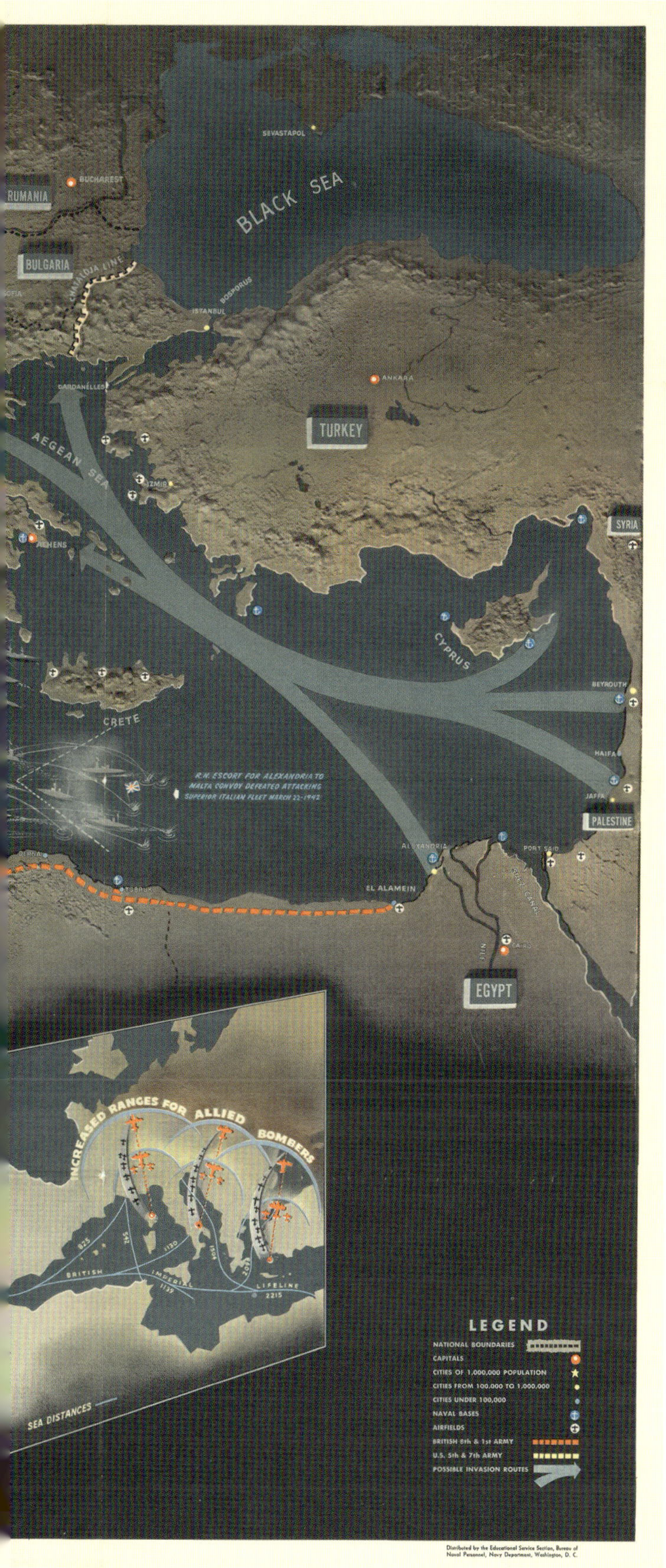

U.S. Navy, Bureau of Naval Personnel: The Mediterranean

lithographed map, U.S. Government Printing Office for The Educational Services Section, Navy Dept., Washington, D.C., 1944,
57 x 38 in. (145 x 96 cm)

This is a fine example of the care and effort the Allies devoted to motivating their vast numbers of disparate fighting men through education, rather than the Axis Powers cruder reliance on indoctrination.

One of the principal aims of the Educational Services Section of the Navy was to inform the American service personnel about the background and context of the battles in which they were fighting. In addition to illustrations of contemporary events, the lower part of the map is given over to educational facts about the region for the benefit of the thousands of soldiers and sailors in this theatre, the majority of whom would never have previously left North America. Particular emphasis is placed on the long U.S. involvement in the region and how the origins of many Americans can be found in and around the Mediterranean.

The Mediterranean was one of the most important, complex, and intensively fought over theatres of the Second World War. All the principal combatants in the European arena became involved in a life and death struggle upon and beneath its waters, in the air above and on its bordering lands.

The key to controlling the Mediterranean had always been naval superiority but for the first time the importance of air power became increasingly significant. Warships and the vital slow-moving supply ships connecting Europe, Africa and the Middle East were now acutely vulnerable to aerial attack. With both sides desperate for suitable naval harbours and airbases, the battle raged back and forth and control of Rhodes, Crete, Malta, Sicily, Sardinia, Corsica, the Balearics, and many other islands became one of the main strategic objectives for the opposing commanders.

Most of the illustrations relate to sea battles, defences of or attacks on supply convoys, bombardments of coastal cities and defences or invasions from the sea. The 850 vessels employed in Operation Torch, the Allied invasion of French North Africa, are shown streaming through the Straits of Gibraltar. A note on the map relating to Operation Husky, the subsequent invasion of Sicily reads: "The largest convoy ever assembled – 2000 ships, carried the Allied invasion forces to Sicily, July 9-12, 1943." This was of course only the case until 6 June 1944 when the massive invasion fleet assembled for Operation Overlord far exceeded that number.[205]

[205] *Operation Neptune* for D-Day involved 6,939 vessels and was the largest naval force in history. It comprised 1,213 naval combat ships, 4,126 landing ships and landing craft, 736 ancillary craft and 864 merchant vessels manned by 195,700 naval personnel. www.ddaymuseum.co.uk/d-day/

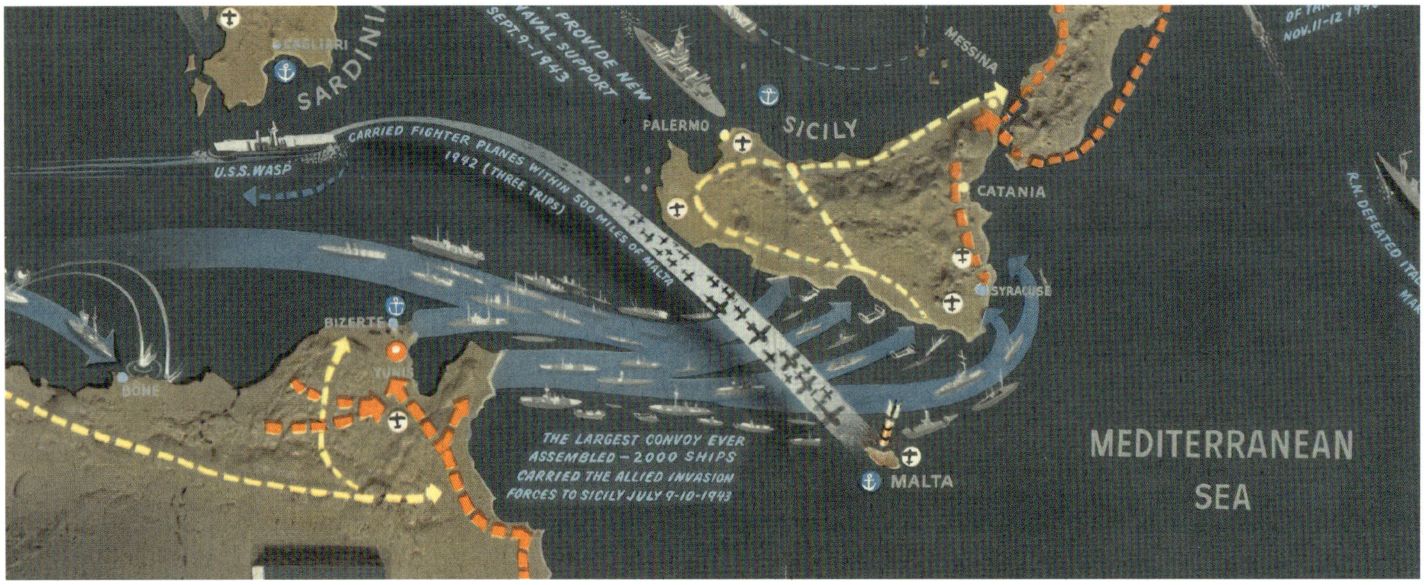

The most strategically important island in the Mediterranean prior to the invasion of Sicily was undoubtedly Malta. Initially regarded as indefensible by the British, already hard-pressed in the Battle of Britain, the island had virtually no air defences and was only protected by six elderly biplanes. The Italian air force dropped their first bombs on Malta within hours of Mussolini declaring war.[206]

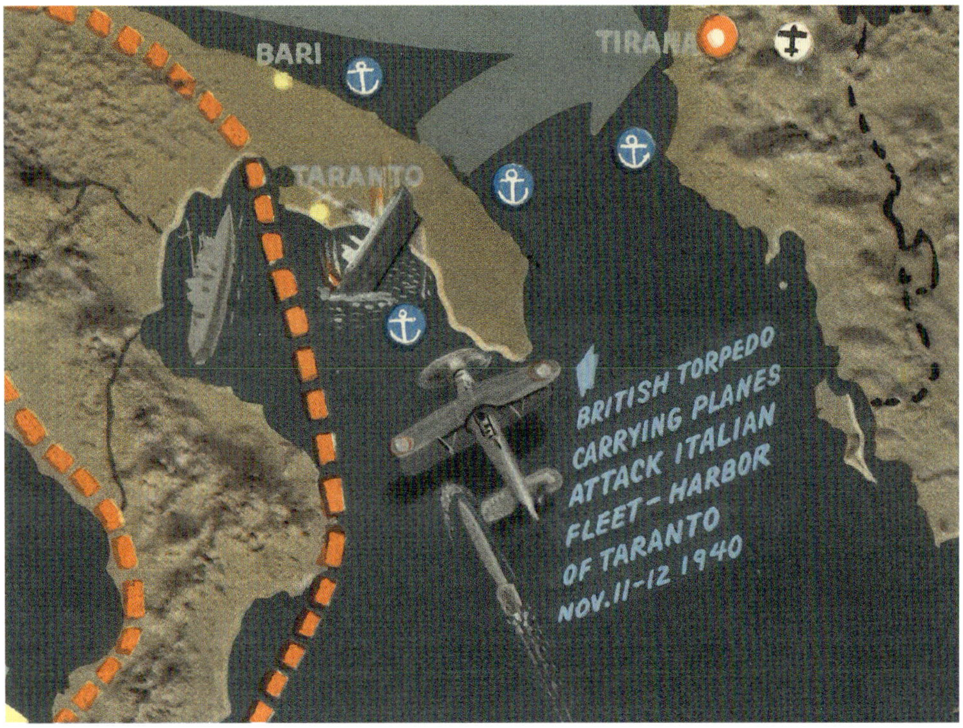

It was only Italy's awe of the fabled might of the Royal Navy which prevented an immediate invasion of the virtually unprotected island. This awe was reinforced by the shock sinking of three of their battleships in one night by British carrier borne torpedo bombers at the Battle of Taranto.[207] The ability of Allied ships and planes based on Malta to cut the vital supply lines to North Africa, first of the Italian Army and then Rommel's Afrika Korps, now became absolutely crucial to the entire war in the Mediterranean. In 1941 Rommel declared that "Without Malta the Axis will end by losing control of North Africa!"[208] Malta became the focus of a grim race between Axis

[206] Beevor, Antony, 2012, *The Second World War*, London, p. 146.

[207] The attack against the Italian fleet moored in the Taranto harbour by twelve obsolete Swordfish torpedo bombers launched from the aircraft carrier, *Illustrious*, was the first all-aerial ship to ship combat in history. It resulted in the disabling of half of Italy's capital ships, including the battleships *Cavour*, *Duilio* and *Littorio*, for the loss of only two aircraft and gave the RN naval superiority in the Mediterranean for the remainder of the war. Beevor, Antony, 2012, *The Second World War*, London, p. 149.

[208] Taylor, A. J. P. & Mayer, S. L. (eds.), 1974, *A History of World War Two*. London, p. 182.

bombing raids determined to destroy all Allied land-based offensive and defensive capabilities and the Allies desperate attempts to reinforce and strengthen the island's fighter air cover.[209] Because of the distances involved, the Allies were forced to transport the planes by air-craft carrier through the Straits of Gibraltar to within 500 miles of the island before launching them. The American carrier, *Wasp* transported two (not three as shown on the map) such vital deliveries enabling Winston Churchill to send a message to the ship, "Many thanks to you all for the timely help. Who said a wasp couldn't sting twice?" By August 1942, despite over 3,000 separate Axis bombing raids against the little island, the Allies had achieved air superiority and by November the Siege of Malta was effectively lifted.[210] The subsequent Allied control over the Mediterranean sea-lanes helped ensure the inevitable defeat of the Afrika Korps starved of supplies and trapped in North Africa.[211] The Allies had sunk so many Axis ships during 1941, that the veterans of the Afrika Korps began to refer to the Mediterranean as the "German Swimming Pool".[212]

Another interesting illustration is of the Battle of Cape Matapan. In the course of this three-day battle from 27 to 29 March 1941, the Italians lost three heavy cruisers and over 2,300 casualties, their worst ever naval loss of life. The Royal Navy lost only three airmen, when a torpedo bomber was shot down. The extraordinary disparity of casualties was due to the Italian navy's lack of radar. This rendered the ships of the Regia Marina effectively blind at night. On the night of 28 March, the British battleships, *Valiant*[213], *Barham* and *Warspite* were able to sail unnoticed to within point blank range of the Italian heavy cruiser fleet before opening fire.[214] The Italians, believing a night attack was impossible, had not even unshipped their guns. This proved to be the last fleet action to be fought by the Royal Navy.[215]

[209] Beevor, Antony, 2012, *The Second World War*, London, p. 155

[210] As recognition of the bravery with which the Maltese inhabitants endured over two years of bombing, the entire island was awarded the George Cross. This was the first ever collective award for Britain's highest civilian gallantry medal and the cross is still featured on Malta's flag.

[211] Beevor, Antony, 2012, *The Second World War*, London, p. 224

[212] Heckman, Wolf, 1981, *Rommel's War in Africa*, New York, p. 157

[213] Prince Philip, the future Duke of Edinburgh, then a 20-year-old midshipman in HMS *Valiant*, commanded the ship's searchlights and was mentioned in dispatches for his conduct during the battle.

[214] Kaplan, Philip, 2014, *World War Two at Sea: The Last Battleships*, , p. 116; Seth, Robert, 1960, *Two Fleets Surprised. The Story of the Battle of Cape Matapan, Mediterranean, March, 1941*, London.

[215] The *Battle of North Cape* on 29 December 1943 during which the *Scharnhorst* was sunk is considered the last battle between big gun capital ships but is not counted as a 'fleet action' as the *Scharnhorst* was without escorts and therefore not part of a fleet.

OUR NAVY BREAKS THE U-BOAT 'SCOURGE ON THE ALLIES' SUPPLY LINES WITH DESTROYERS, DESTROYER
2 MAY 23–THE BISMARCK WAS SIGHTED BY THE BRITISH CRUISERS SUFFOLK AND NORFOLK WHO CONTINUED TO SHADOW HER THROUGHOUT THE NIGHT.
REYKJAVIK
ICELAND
JULY 7, 1941–U.S. NAVAL FORCES OCCUPY ICELAND
MAY 24 – AFTER A
HARP ENGAGEMENT IN
HICH SHE SANK H.M.S. HOOD
ND DAMAGED H.M.S. PRINCE OF WALES,
HE CRIPPLED BISMARCK ELUDED
EARCHING PLANES AND SHIPS
HILE FRANTICALLY TRYING TO
EACH THE FRENCH COAST.
GERMAN INVASION OF NORWAY APRIL
BERGEN
SCAPA FLOW
OCT. 14, 1939 H.M.S. ROYAL OAK WAS SUNK BY TORPEDO FROM GERMAN SUB.
SCOTLAND
LONDONDERRY
GLASGOW
EDINBURGH
NORTH SEA
IRISH SEA
DUBLIN
EIRE
LIVERPOOL
ENGLAND
HULL
THE R.A.F. AND THE A.A.F. CONTROL THE INDUSTRIAL HEART OF EUROPE FROM ENGLAND. THE WORLD'S MOST POWERFUL AIR BASE
AMSTERDAM
ROTTERDAM
DUNKIRK
BELGIUM
AACHEN
PORTSMOUTH
ENGLISH CHANNEL
CHERBOURG
ROUEN
PARIS
BREST
LORIENT
ST. NAZAIRE
LOIRE R.
SEINE R.
FRANCE
MONTLUCON
MAY 26 – CATALINAS
OCATED THE BISMARCK
AND A SWORDFISH
ORPEDO WRECKED
HER STEERING GEAR.
THAT NIGHT THREE
ORPEDO HITS WERE
SCORED BY CLOSING-
N DESTROYERS.
5 MAY 27 – MAIN BRITISH FORCE CAME UP AND WITHIN AN HOUR THE CORNERED BISMARCK WAS A BLAZING WRECK. REFUSING TO SURRENDER, SHE WAS FINALLY SUNK BY TORPEDOES FROM THE DORSETSHIRE.
ATLANTIC OCEAN
TO MEDITERRANEAN PORTS
DEC. 28, 1943 – TWO U.S. NAVAL FLYERS SPOTTED A FLOTILLA OF GERMAN DESTROYERS AND SHADOWED IT UNTIL A BRITISH FORCE ARRIVED. THREE NAZI DESTROYERS WERE SUNK AND SEVERAL DAMAGED.
BAY OF BISCAY

U.S. Navy, Bureau of Naval Personnel: World War 2 in the North Sea Area

U.S. Navy, Bureau of Naval Personnel: World War 2 in the North Sea Area

lithographed map, U.S. Government Printing Office for The Educational Services Section, Navy Dept., Washington, D.C., 1944, 57 x 38 in. (145 x 96 cm)

Winston Churchill wrote that "The Battle of the Atlantic was the dominating factor all through the war…" and this is the focus of this large pictorial map of the North Sea and North Western Europe.[216] It highlights the all-important convoy routes bringing materiel and men to Britain, North Africa and Russia. A comment above the path of the convoys declares "Our navy breaks the U-boat scourge on the Allies' supply lines with destroyers, destroyer escorts and escort carriers". The number of ships lost had gone from 1661 (1159 by U-boat) in 1942, the height of the U-boat menace, to 597 (463 by U-boat) in 1943.

Other significant features shown are the bombing targets in occupied Europe within reach from the airfields based in south-east England, and the German invasion routes to Norway.

This map and the others in the series were created as training aids and distributed by the Educational Services Section, Bureau of Naval Personnel. There is explanatory text across the map, detailing key moments during the war so far, particularly those of naval interest often illustrated with vivid vignettes.

One of the earliest incidents of the war at sea and one of the most shocking for the British, was the sinking of the *Royal Oak* by a U-boat while at anchor in Scapa Flow on 14 October 1939. Completed in 1916 and first seeing action in the Battle of Jutland, the *Royal Oak's* lack of speed made her unsuited for front-line activity in the Second World War. She was anchored as a floating anti-aircraft battery when the German submarine, *U-47* under the command of Günther Prien, managed to sneak into the supposedly impenetrable British naval base and launched three salvoes of torpedoes, sinking the battleship within minutes with the loss of over 800 of her crew of 1,200.[217] The propaganda value of this blow against the mighty British Navy was immense. Günther Prien, the commander of *U-47*, became a hero overnight and was the first submarine officer to be awarded the Knight's Cross of the Iron Cross.[218]

Also illustrated in a series of vignettes is the epic story of the breakout and eventual sinking of the *Bismarck*. The *Bismarck* and her sister ship, the *Tirpitz*, were the largest battleships built by Germany and at 793 feet (241m) and 41,000 tons the largest built in Europe.[219] Intending to fulfil her role as a commerce raider, on 20 May 1941 the *Bismarck* along with the heavy cruiser, the *Prinz Eugen* and accompanied by her attendant escorts, slipped out of the Baltic into the North Sea. Alerted by friendly elements in the Swedish Navy of the intended sortie and backed up by German radio traffic intercepted by Bletchley Park, the British Admiralty mobilised the Home Fleet at Scapa Flow to intercept the raiders.

[216] Churchill, Winston, 1951, Closing the Ring, *History of the Second World War, vol. 5*, London.

[217] Beevor, Antony, 2012, *The Second World War*, London, pp. 40, 183.

[218] Günther Prien, the first German submarine ace of the Second World War, would go on to sink nearly 200,000 tons of Allied shipping before the *U-47* went missing in action, probably sunk by depth charges from *Wolverine*, in March 1941.

[219] This was only exceeded at 814 feet (248m) and 44,500 tons by the British fast battleship, *Vanguard*. Ordered in 1941 but unfinished at the end of the war, she was already obsolete by the time she was finally commissioned in 1946. Eventually scrapped in 1960, she was the last battleship ever to be launched.

GERMAN INVASION OF NORWAY APRIL 1940
BERGEN
STAVANGER
SCAPA FLOW
OCT. 14, 1939
H.M.S. ROYAL OAK WAS
SUNK BY TORPEDO
FROM GERMAN SUB.
SCOTLAND

BREST
LORIENT
ATLANTIC
OCEAN
DEC. 28, 1943 – TWO U.S. NAVAL
FLYERS SPOTTED A FLOTILLA
OF GERMAN DESTROYERS AND
SHADOWED IT UNTIL A BRITISH
FORCE ARRIVED. THREE NAZI
DESTROYERS WERE SUNK AND
SEVERAL DAMAGED.
BAY OF BISCAY

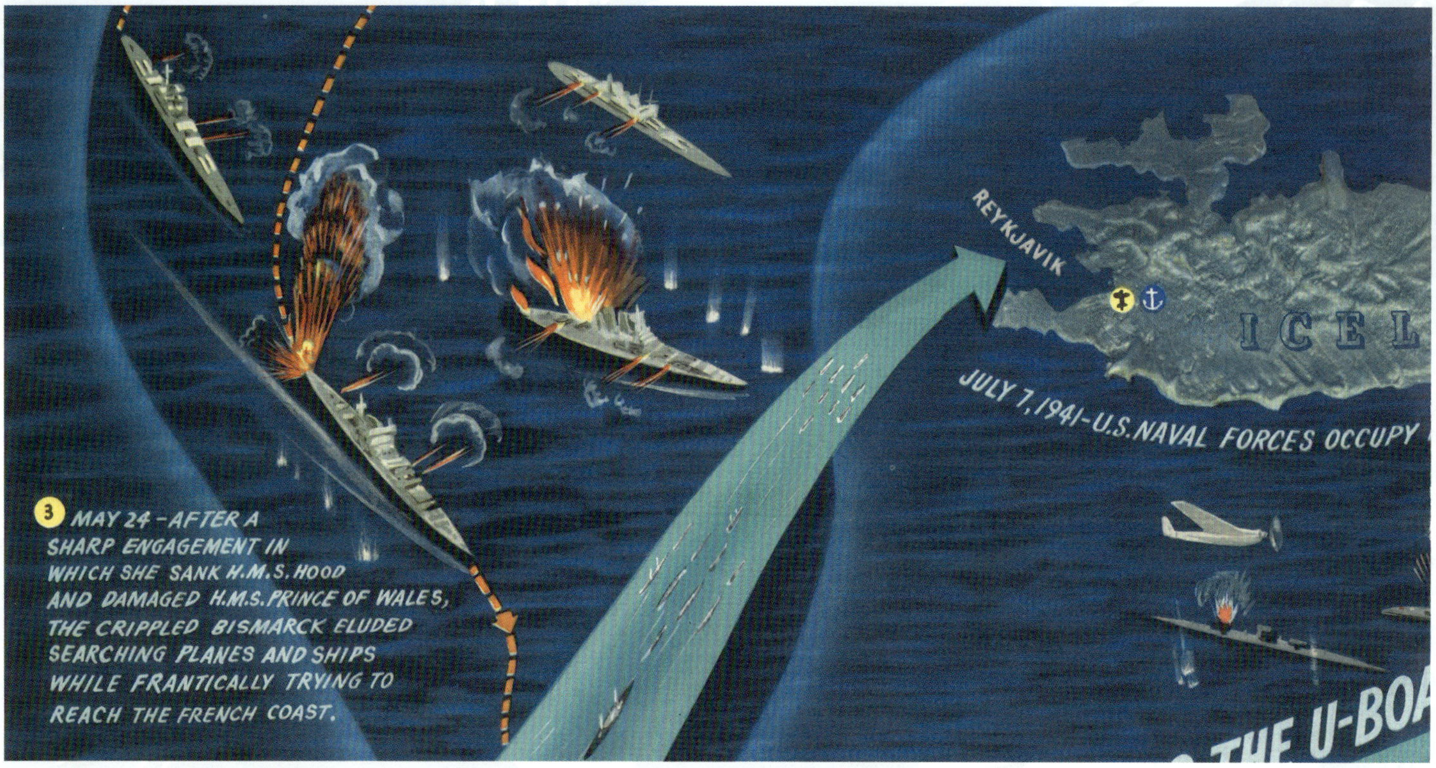

On 23 May, Admiral Lütjens, commander of the *Bismarck*, ordered his ships to increase their speed to 27 knots (50km/h; 31mi/h) and make the dash into the North Atlantic through the choke point of the Denmark Strait.[220] The following day they were sighted by the British cruisers *Suffolk* and *Norfolk* who continued to shadow them through the night while relaying their position to the nearest battleships, *Hood* and *Prince of Wales*.[221] At dawn on 24 May, the ships were within range of each other and after a ten-minute exchange of salvos, the *Hood* was destroyed by a massive explosion, probably caused by a plunging shell from the *Bismarck* penetrating her deck and igniting the aft magazine. Of the 1,418 men on board, there were only three survivors. Despite suffering damage from three hits from the *Prince of Wales*, the *Bismarck* was still capable of reaching 27 or 28 knots, the maximum speed of the pursuing British battleships, and the decision was taken to head for Saint-Nazaire in occupied France for repairs. In the desperate chase that followed, the *Bismarck* managed to lose her pursuers and get within one day's sailing of the coastal zone where she could be protected by the Luftwaffe and U-boat fleet. Once more it was the codebreakers at Bletchley Park, decrypting the German radio traffic, who estimated the battleship's likely heading. On 26 May a Catalina Flying Boat spotted the battleship and was able to transmit her exact position to 'Force H', led by the aircraft carrier *Ark Royal* heading up from Gibraltar at full speed. While still 60 miles from the *Bismarck*, the *Ark Royal* launched her torpedo bombers and Swordfish bi-planes. During a second sortie by 15 Swordfish, two torpedoes struck home, one disabling the *Bismarck's* port rudder forcing her to steam in a great circle. This allowed the chasing force, the battleships, *King George V* and the *Rodney,* supported by the heavy cruisers, the *Norfolk* and the *Dorsetshire* to catch up with the stricken ship. At dawn on 27 May the fleet led by the *King George V* closed on the *Bismarck* and within an hour during which the two British battle ships fired over 700 main battery shells, the pride of the German navy was a blazing helpless wreck. Despite the deaths of her commanders, she still refused to surrender and was finally sunk by torpedoes from the *Dorsetshire*.[222]

[220] Confusingly the Denmark Strait is actually the strait between Iceland and Greenland.

[221] Built in 1920, *Hood* was technically a battle cruiser, not having the armour protection of a modern battleship such as the Bismarck. The Admiralty was aware of its flaws and it was due for urgent upgrading which was indefinitely delayed by the start of the war.

[222] Beevor, Antony, 2012, *The Second World War*, London, pp. 184-185. The 1960's film, *Sink the Bismarck!,* starring Kenneth Moore, while generally accurate, had a couple of interesting omissions. Firstly the contribution of the Bletchley Park codebreakers was left out entirely as even in 1960 it was classified as top secret and secondly, the nationality of the American Catalina pilot was glossed over, the U.S. being officially still neutral in May 1941.

The destruction of another much feared German battleship, the *Scharnhorst,* is also shown. During her career as a commerce raider, the *Scharnhorst* sank thousands of tons of merchant ships as well as the aircraft carrier, *Glorious* and her two destroyer escorts.[223] The Germans were unaware that their radio traffic was being intercepted, and when the *Scharnhorst* was dispatched to attack a North Atlantic convoy bound for Russia, the battleship *Duke of York*, and the cruisers *Norfolk*, *Sheffield*, and *Belfast* were waiting for her. On 26 December 1943, after a fierce exchange of main battery salvoes, the *Scharnhorst* was at last sunk by a combination of shellfire and torpedoes. Of her crew nearly two thousand, only 36 survivors were pulled from the icy waters.[224] The Battle of the North Cape, as it came to be known, was the last battle between big gun capital ships of the war.

[223] During the engagement with *Glorious* on 8 June 1940, the *Scharnhorst*, at a distance of 15 miles, achieved one of the longest range ship-to-ship big gun hits in history.

[224] Beevor, Antony, 2012, *The Second World War*, London, p. 514.

Keely (Pat): Indie Moet Vrij! Werkt en Vecht Ervoor!

Keely (Pat): Indie Moet Vrij! Werkt en Vecht Ervoor!
[The Indies shall be Freed! Work to Achieve This!]

lithographed map, James Haworth & Brother Ltd., London, 1944, 19 x 29 in. (48 x 74 cm)

The genesis for this poster printed in London for the Dutch Government-in-exile lies in the determination of all the colonial powers, unless irrevocably on the 'wrong side' to reclaim their pre-war overseas possessions.

The arresting image of the malevolent octopus emerging from the rising sun, with its long tentacles grasping the East Indies, represents the stranglehold which Japan had gained over the former Dutch colonies. According to the Imperial War Museum, James Haworth & Brother Ltd. printed 10,000 copies of this poster on behalf of the London-based Netherlands Government.[225]

While the propaganda message of the map is clear, the exact purpose of the poster and whether indeed it was ever actually distributed is more problematic. It may be that when this was commissioned and printed in 1944, it was in anticipation of the imminent Allied liberation of the Dutch East Indies. The intention would have been to distribute it to the re-emerging Dutch colonial authorities and the hopefully grateful peoples of the newly liberated territories. If this was indeed the plan it turned out to be seriously wrong on both counts. Firstly the Americans decided to bypass the Dutch East Indies and push straight on towards Japan itself. Indonesia was to remain in Japanese hands right up until the end of the war. Secondly the Indonesians themselves having now experienced subjugation from both the Dutch and the Japanese and having, at least at first, preferred the latter, were determined upon their own independence which they declared within five days of the Japanese surrender.

Patrick Cokayne Keely was a designer of posters, press advertisements and trade matter for the London Underground, the General Post Office and Southern Railways. Keely's bold use of colour combined with striking symbols produced some of the most powerful images of the war. He was a member of both the Society of Industrial Artists and the Alliance Graphique Internationale.[226] During the war, Keely also produced a striking series of posters for the Royal Society for the Prevention of Accidents for the Ministry of Labour.[227]

[225] It would appear that the Dutch Government-in-exile were enthusiastic propagandists as over the course of the war they were to order 153,000 posters from James Haworth & Brother alone.

[226] Darracott & Loftus, 1972, *First World War Posters*, Imperial War Museum, London, p. 37.

[227] A campaign for the prevention of accidents during a genocidal war may seem pretty futile but in fact the vast number of inexperienced men and women working in war-related industries and especialy the dangers of the blackout had raised the number of civilian casualties to an unprecedented level.

Falter (John), USNR: That's the Spot to Hit! – Give Us the Stuff and We'll Hit It!

Falter (John), USNR: That's the Spot to Hit! – Give Us the Stuff and We'll Hit It!

lithographed map, U.S. Government Printing Office, Washington D.C. for the Official Navy Poster, Industrial Incentive Division, 1944, 28 x 40 in. (71 x 102 cm)

This striking American poster was designed to inspire increased productivity among the hundreds of thousands of workers involved in the war effort.

Admiral Chester W. Nimitz, Commander in Chief, Pacific Fleet is displayed in front of a map of the Pacific Theatre of War, with areas under Japanese control marked in red. A black circle is laid down over the Japanese mainland and concentric circles radiate out from Tokyo.

By 1944 the tide of war in the Pacific was inexorably turning in America's favour. From the Battle of Midway in 1942, American superiority in men and materiel had steadily grown in relation to the Japanese, and the American forces were starting to regain territories lost to the enemy earlier in the war. This aggressive and increasingly confident attitude is vividly reflected in this rare poster designed to encourage the workers making the war supplies back home.

By mid-1944 American forces led by Chester Nimitz were only 300 nautical miles from Mindanao, the largest island in the southern Philippines. Nimitz had already captured the Gilbert Islands and most of the Marshall and Marianas Islands, deliberately bypassing many Japanese army garrisons and outposts who were left isolated and irrelevant.[228] The stage was now set for the next crucial stage in the War in the Pacific: the retaking of the Philippines.[229]

Sometimes overshadowed by the more self-aggrandising character of General MacArthur, Chester Nimitz was one of the principal architects of the Allied victory against Japan. He was born near Fredericksburg, Texas on 24 February 1885, and aged 20 entered the Naval Academy. After graduation he first went to the Far East on board the battleship *Ohio* before transferring to submarines. His climb through the ranks in the navy was mostly done underwater, commanding a number of submarines. Amongst other important commissions, he supervised the construction of the submarine base at Pearl Harbor in 1920. Having served as aide to the Commander in Chief through the 1930s, he was promoted to Rear Admiral in 1938. Ten days after the attack on Pearl Harbor, he was appointed Commander in Chief of the Pacific Fleet with the rank of Full Admiral.[230] The change of command ceremony would normally have been conducted on the bridge of a battleship, but because every battleship had been either sunk or badly damaged, the ceremony, fittingly for Nimitz, took place on the upper deck of a submarine.

Nimitz is shown wearing the shoulder insignia of a Full Admiral, dating it prior to 19 December 1944 when he was promoted to the newly created rank of Fleet Admiral and received his fifth star. He and McArthur were the principal American signatories to the Japanese instrument of surrender, itself signed on the deck of the USS *Missouri* on 2 September 1945.[231]

The artist John Falter was just starting to make a name for himself as an illustrator when the Second World War broke out. Originally from Nebraska, he sold his first art piece in 1930 to the magazine *Liberty*, which paved the way for his many commissions throughout a long career, working for Gulf Oil Company, Four Roses Whisky, *The Saturday Evening Post*, amongst others. In 1943 he enlisted in the U.S. Navy and was rapidly promoted from a chief boatswain's mate to lieutenant, with special responsibility for recruiting and propaganda art.[232]

[228] Beevor, Antony, 2012, *The Second World War*, London, p. 565.

[229] The last Japanese soldier to surrender in the Philippines, Second Lieutent Hiroo Onada was only persuaded to give himself up in 1974. He died in January 2014 aged 91.

[230] Beevor, Antony, 2012, *The Second World War*, London, p. 306.

[231] The Naval History and Heritage Command, at the Washington Navy Yard: www.history.navy.mil/faqs/faq36-4.htm

[232] http://fortmissoulamuseum.org/WWII/detail.php?id=403

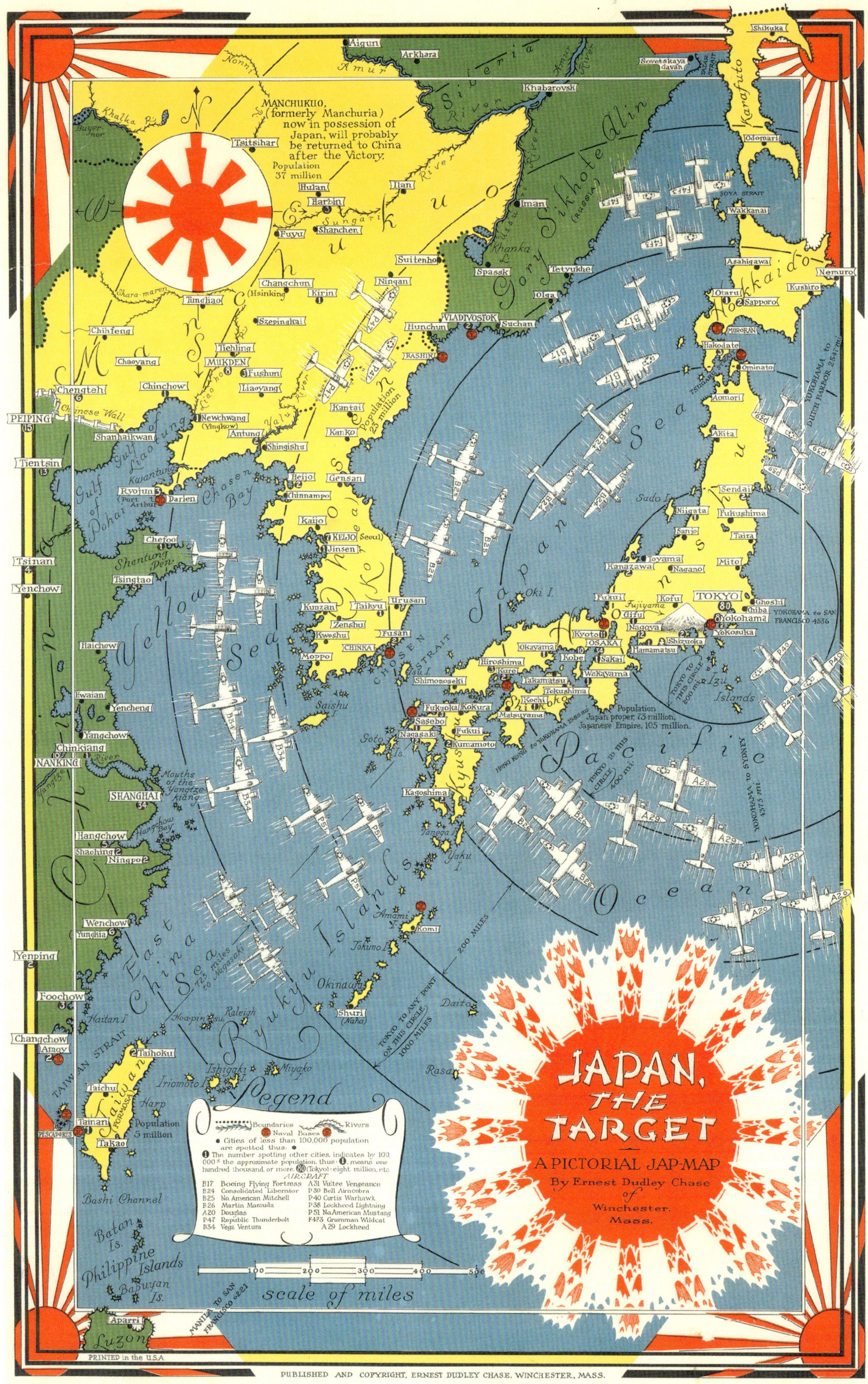

Dudley Chase (Ernest): Japan the Target – A Pictorial Jap Map

Dudley Chase (Ernest): Japan the Target – A Pictorial Jap Map

lithographed map, Ernest Dudley Chase, Winchester, 1944, 15 x 22 in. (39 x 56 cm)

Dating from the closing stages of the war in the Pacific, this is one of the scarcest of Dudley Chase's pictorial maps. Waves of bombers are shown heading for Japan with concentric rings radiating out from Tokyo indicating distance.

This arresting map represents a decisive stage of the war strategically and psychologically for both the Americans and the Japanese. Using their island bases, as well as those in the Philippines and China, the American land-based bombers were at last within striking range of mainland Japan.[233] Each of the planes is labelled on the wing, referencing to an index in the lower left-hand corner. The B-17 Flying Fortress, the Grumman Wildcat and others are all shown. The principal bomber that is missing is the B-29 Superfortress, which was only introduced at the end of 1944, and which because of its long range became the primary instrument for the fire-bombing of Japanese cities towards the end of the war.[234]

Unusually for such an overtly partisan propaganda piece, Dudley Chase has incorporated the enemy's Imperial emblem and flag into the decorative motifs.

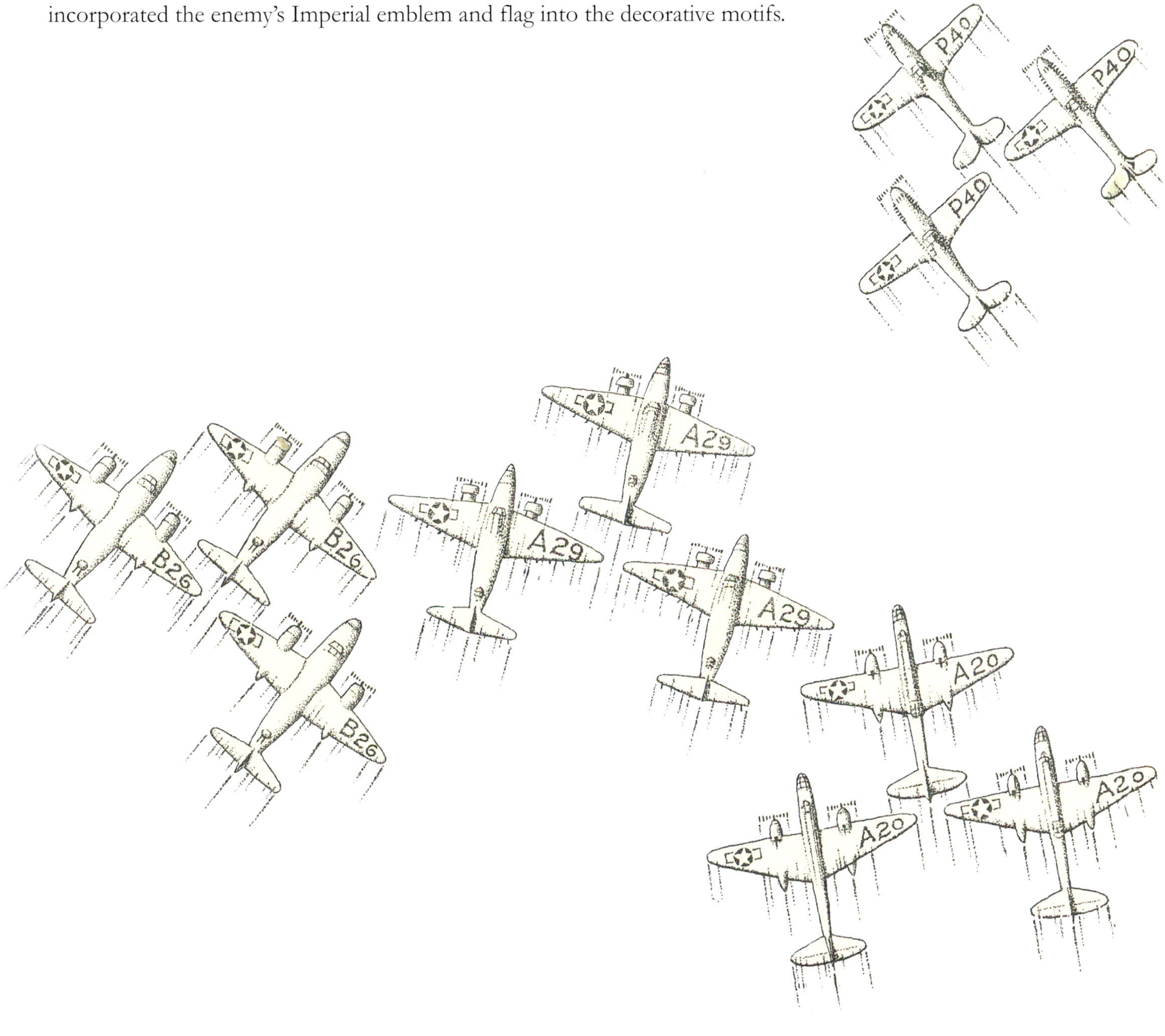

[233] The earlier Doolittle Raid bombing raid on Tokyo with 16 B-25 Mitchell bombers had been launched from carriers.

[234] It was the B-29s, *Enola Gay* and *Bockscar*, that dropped the atomic bombs on Hiroshima and Nagasaki.

Rand McNally & Co.: Map of the United States Military Posts

Rand McNally & Co.: Map of the United States Military Posts

lithographed map, Rand McNally & Company, Chicago, 1944,
26 x 19 in. (66 x 48 cm)

Following on from C.S. Hammond's pictorial map, Rand McNally, the leading American map publisher during the twentieth-century, issued a map celebrating US military might. This Rand McNally piece is simpler in its design, with drawings and symbols added to a standard map of the country, without the decorative border of the earlier map.

The huge number of military bases, chiefly along the Eastern Seaboard, is a vivid reminder of the scale of the U.S. war effort at this crucial period, leading up to the invasion of mainland Europe while simultaneously fighting a vast naval and land war in the Pacific Theatre.

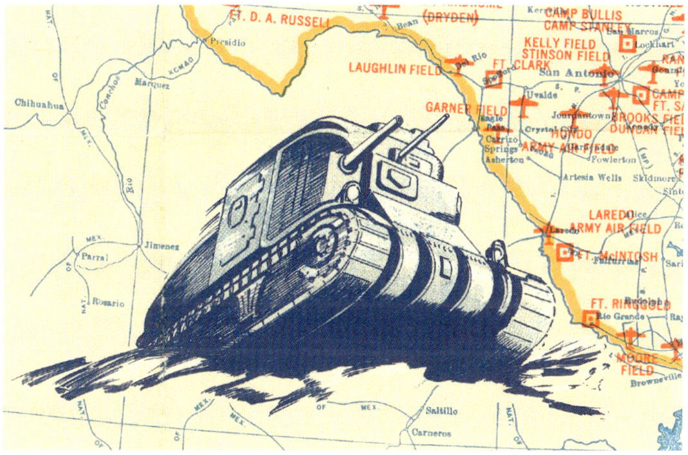

Among the military vignettes of ships, planes and personnel is an interesting image of an M-3 Lee tank. This archaic tank with its high turret profile, hull mounted main gun and a distinct reluctance to go off road, performed surprisingly well in the North African and Burma campaigns.[235] From the end of 1942, it was steadily replaced with the M-4 Sherman and finally declared obsolete in April 1944.[236]

[235] Desperate for medium tanks, The British ordered 1,250 M-3s from the American manufacturers at a cost of $240,000, the total sum of British government funds in the U.S.. Under lend-lease a further 1,400 were supplied to Soviet Russia where they were dismissively described as "a coffin for six men!".

[236] The early Shermans were also regarded with little affection with the British calling them "Ronsons" after the Ronson cigarette lighter slogan which went "It lights first time, every time". The Germans contemptuously referred to them as "Tommycookers"!

Anon: Това би могла да очаква ЕВРОПА отъ болшевизма

Anon: Това би могла да очаква ЕВРОПА отъ болшевизма [This is what can be expected if Europe succumbs to Bolshevism]

lithographed map, [1944], 33 x 23 in. (84 x 58 cm)

This exceedingly rare piece of savage anti-Bolshevik propaganda was issued as the Russian Army approached ever closer to the borders of the pro-German Bulgarian puppet state in the latter part of the Second World War.

Bulgaria probably had the most anomalous position of all the Axis powers' satellite states. At the start of the war the government of the Kingdom of Bulgaria was determined to remain neutral. It was not until 1 March 1941 that the threat of a German invasion and the tempting possibility of seizing territory in the Balkans persuaded them to join the Axis Bloc. They reluctantly declared war against Britain and the U.S. but held back from declaring war on Russia even after the ending of the Molotov-Ribbentrop Pact.[237]

When Germany invaded Russia in June 1941 a substantial Communist-led resistance movement, the Fatherland Front, mobilised against the pro-German Bulgarian government.

[237] Beevor, Antony, 2012, *The Second World War*, London, pp. 154-156.

Under German pressure, the Bulgarian government enacted a number of anti-Semitic laws effectively stripping the native Jewish population of almost all rights and their citizenship. But when the Germans demanded that the entire Bulgarian Jewish population of 48,000 should be shipped to the concentration camps, something unprecedented and remarkable happened. The military government, supported by Tsar Boris III and the Orthodox Church and encouraged by widespread public support, first prevaricated then flatly refused to allow any of the Jews from Bulgaria to be deported. Virtually the entire Jewish population of Bulgaria survived the war.[238]

The stark warning of the savagery that would accompany the successful conquest of Europe by Soviet Communism was a frequent message aimed at the conquered and satellite nations under Nazi domination. The emphasis in this piece is targeted towards Soviet repression of organised religion and depicts the destruction of churches, both Catholic and Orthodox, and the murder of nuns and priests. In Italy two priests are hanging from a gibbet while in Anatolia, Turkey, a mosque is in flames. In Britain, the Lord Chancellor is being brutally hacked down in front of a burning Houses of Parliament.

Lacking any imprint, it is impossible to tell whether this propaganda piece was actually printed in Germany to try and garner wavering Bulgarian public support in the face of the Soviet successes in the east, or printed locally to bolster the royalist government. Either way, it proved ineffectual as on 9 September 1944, with the triumphant Russian army already across its borders, a Fatherland Front-led coup deposed the government, welcomed the Russians as liberators and immediately declared war on Nazi Germany.[239]

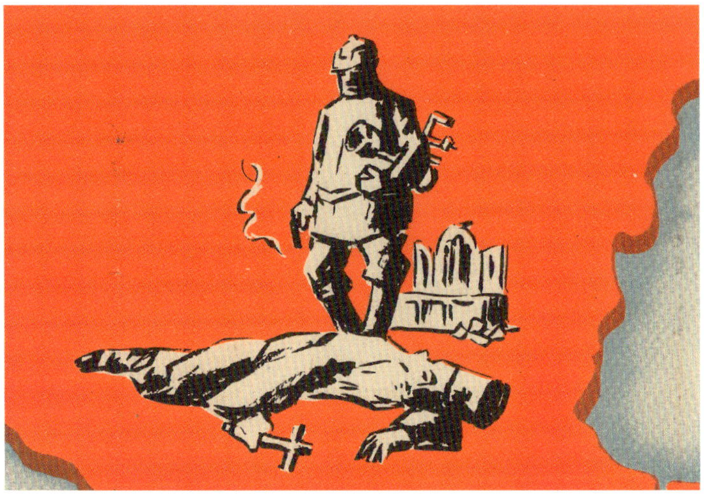

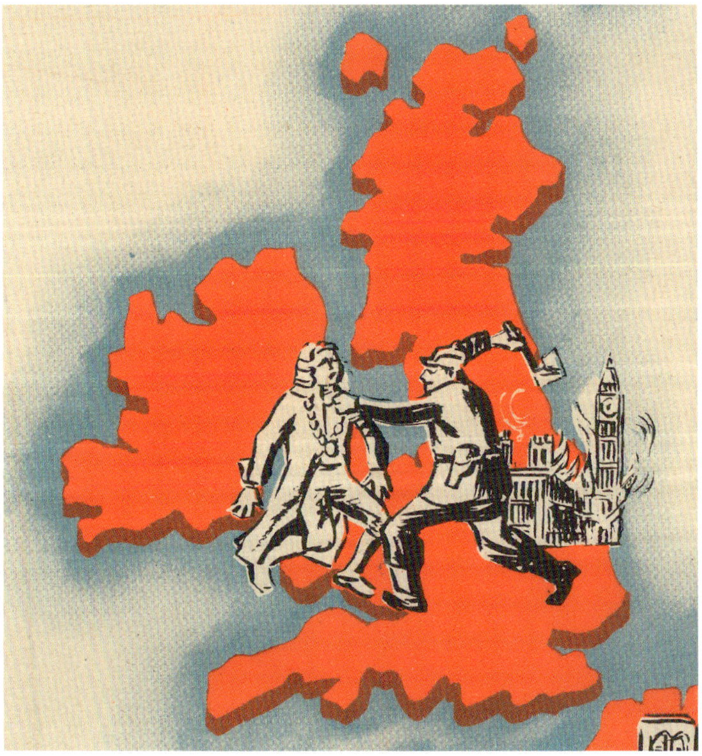

[238] The extraordinary story behind the salvation of the Bulgarian Jews only came to light at the end of the 1980s as it was suppressed by the Soviet government after the war because of the active involvement of the former government, the Church and Tsar Boris III, all considered enemies of communism. See: Bar-Zohar, Michael, *Beyond Hitler's Grasp: The Heroic Rescue of Bulgaria's Jews*, Avon, MA., 2001.

[239] Beevor, Antony, 2012, *The Second World War*, London, p. 636.

Anon: A L'Est – L'URSS Indomptable, serre les Kamerads

coloured lithographed map, [France], 1945, 3.5 x 5.5 in. (79 x 14 cm)

Anon: A L'Est – L'URSS Indomptable, serre les Kamerads

"To the East – The Invincible USSR, tightens its grip on the Kamerads".[240]

This French postcard, published after the Liberation, celebrates the Russian successes against the Germans on the Eastern Front, a striking contrast to the fanatically anti-communist attitude of Vichy France.

[240] The derisive term "Les Kamerads" for Germans originated from the First World War and is derived from the supposed cry of "Kamerads! Kamerads!" of surrendering German soldiers.

Furman (Larry): [Europe] - H. M. E. Pendleton, Colonel, Cav. Commanding, European Civil Affairs Division

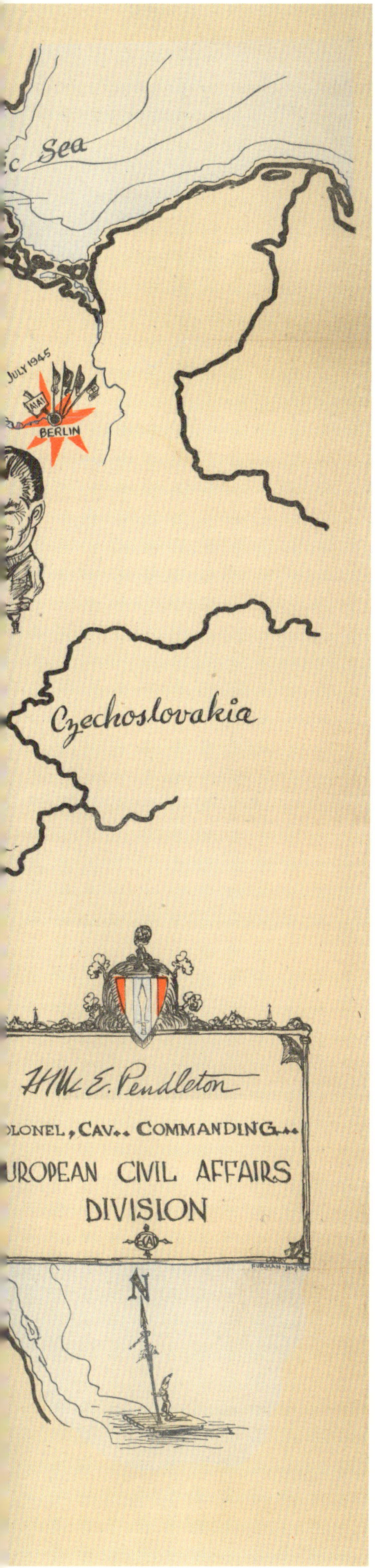

Furman (Larry): [Europe] - H. M. E. Pendleton, Colonel, Cav. Commanding, European Civil Affairs Division

lithographed map, [Germany?], July 1945, 21 x 18 in. (53 x 46 cm)

This exceptional map celebrates the U.S. Civil Affairs Division, one of the most significant and indeed powerful organisations created during the war whose vital work does not always get the credit it deserves.

At the start of the war, Roosevelt felt there should be a total separation between the military, whose job was to prosecute the war, and the civilian authorities who should manage infrastructure and people in the liberated and conquered territories. The weakness of this division of labour became immediately apparent after the start of Operation Husky and the invasion of North Africa. The civilian agencies were completely inadequate, both in manpower and expertise, in rebuilding the ruined infrastructure as well as looking after the huge numbers of destitute and often starving people in the captured territories. This rapidly turned into a logistical and political disaster with General Eisenhower complaining bitterly that "I am having as much trouble with civilian forces behind aiding us as I am with the enemy in front of us."[241]

Recognising that only the army had the resources to deal with the enormous task of reconstruction and controlling the now leaderless and sometimes hostile populations, the Civil Affairs Division (CAD) was created on 1 March 1943, placing all civilian planning under the auspices of the War Department.

The civil affairs missions in Germany and Japan during and immediately after the Second World War, saw the army go far beyond its traditional role, clearing roads, repairing sewage lines or assisting refugees. By the end of the war, CAD would be responsible for the wellbeing of over 80 million civilians.[242]

Germany was the greatest challenge; while the experience after the First World War had taught the planners that while a weak and degraded Germany was a breeding ground for resentment, two world wars had shown that an independently strong nation was hardly ideal either. The strategy adopted by the U.S. was to maintain control over the different departments of local government rather than govern the civilian population directly.[243] Paired with the aid in economic rebuilding, this proved generally successful. However, direct control became necessary after the final collapse of the Third Reich and the de-Nazification purges of 1945, which prohibited over 25 per cent of the German population from performing civil services, or indeed any work except manual labour, and which left virtually no functioning government agencies.[244]

[241] Zaalberg, Thijs Brocades, 2006, *Soldiers and Civil Power: Supporting or Substituting Civil Authorities in Modern Peace Operations*, Amsterdam, p. 26.

[242] Dobbins, James; Poole, Michele A.; Long, Austin & Runkle, Benjamin, 2008, *After the War: Nation Building from FDR to George W. Bush*, Santa Monica, CA., p. 28.

[243] Oehrig, Cristen, 2009, *Civil Affairs in World War II*, Center for Strategic and International Studies, Washington, DC., p. 5.

[244] Dobbins, James; Poole, Michele A.; Long, Austin & Runkle, Benjamin, 2008, *After the War: Nation Building from FDR to George W. Bush*, Santa Monica, CA., p. 28.

The Civil Affairs Division was also involved in the task of returning the vast quantity of Nazi looted art and artefacts to their rightful owners whether individuals or states. The so-called 'Monuments Men' in *The Monuments, Fine Arts and Archives Program* nominally served in CAD but in practice usually came under the direct command of the SHAEF (Supreme Headquarters Allied Expeditionary Force) commander in the field.

As well as geographical places of interest, the map is liberally illustrated with vignettes of one of the most persistent and intractable problems to plague the military and civilian authorities: how to stop or at least curtail excessive "fraternisation" with the natives. In France, one young lady is shown dropping her handkerchief in order to lure a GI through a door labelled "Off Limits", another plies a drunken soldier with wine, while a third is being chased around the Eiffel Tower.

Smiling inscrutably, Joseph Stalin dominates the Russian Zone while the American Zone bears a sign saying "Under New Management, U.S.A. Mil. Govt."

This map also obliquely commemorates one of the most extraordinary tales of German derring-do of the later part of the war. Alongside the landing zone in Normandy, there is a tiny vignette of three American prisoners labelled "3 ECAD Officers captured in the Granville Raid". In December 1944, four German paratroopers and a naval cadet managed to escape from Granville PoW camp in Allied Occupied France, stole an American landing craft and sailed it to the German-held Channel Islands. Once there, they reported to Admiral Friedrich Hüffmeier, the Governor of the Channel Islands, that several freighters were discharging coal in Granville Harbour. Badly in need of a morale-boosting victory and desperately short of coal, Hüffmeier decided to send a force under the command of Kapitänleutnant Carl-Friedrich Mohr to capture the ships. Mohr, leading a motley flotilla of thirteen vessels, including seven minesweepers, three armed barges and an ocean-going tug, entered Granville harbour and successfully captured a coal freighter. Despite coming under sustained American small arms fire, a German shore party held the harbour, burning and blowing up several of the dock facilities and mining a number of ships grounded by the low tide. They also freed over fifty German PoWs. Another group of twenty-five soldiers launched a diversionary attack on the Hotel des Bains and Hotel Normandie and, after meeting little resistance, captured nine senior American officers reportedly with help of a French female member of staff. It was presumably here that the ECAD officers were picked up.[245] After his triumphant return to the Channel Islands, Mohr received the Knight's Cross. Admiral Canaris, head of the Abwehr, German military intelligence, ruled out any further raids so close to the end of the war he already knew was lost.[246]

The Civil Affairs Division, now under the umbrella term of Civil-Military Operations (CMO), has seen constant service in every field of conflict involving American troops from Korea, the Balkans and the Middle East and more recently, in Afghanistan.

[245] Kaufmann, J. E. & Kaufmann, H. W., 2010, *The American GI in Europe in World War II*, Mechanicsburg, PA., p. 366; Sanders, Paul, 2005, *The British Channel Islands under German Occupation, 1940-1945*, Jersey, p. 181; Stasi, Jean-Charles, 2015, *L'Incroyable Raid de Granville*, Bayeux.

[246] Only weeks before the end of the war, Admiral Wilhelm Canaris was hanged at Flossenbürg concentration camp for his role in a number of plots to overthrow or kill Hitler.

BREMEN
"V-E" DAY
8 MAY 1945
BERLIN
"JOE"
SHING ON THE SIEGFRIED LINE
MUST BE A WAC!
KÖLN
VERBOTEN?
10 MAY 1945
FINAL HQ. ECAD
ATERNIZATION
BAD HOMBURG
WIESBADEN
FRANKFURT
UNDER NEW MANAGEMENT
-U.S.A.-
MIL. GOVT.
WESTERN DISTRICT
U.S. ZONE - HQ. AT
HEIDELBERG
AARBRUCKEN
Czecho

Channel
CHATEAU MONT EPINQUET
FWD ECH HQ ECAD
7 JUNE
D+1
1944
CIVIL AFFAIRS
CHERBOURG
3 ECAD OFFICERS CAPTURED AT GRANVILLE RAID
DIEPPE
LE HÂVRE
ST LO
CAEN

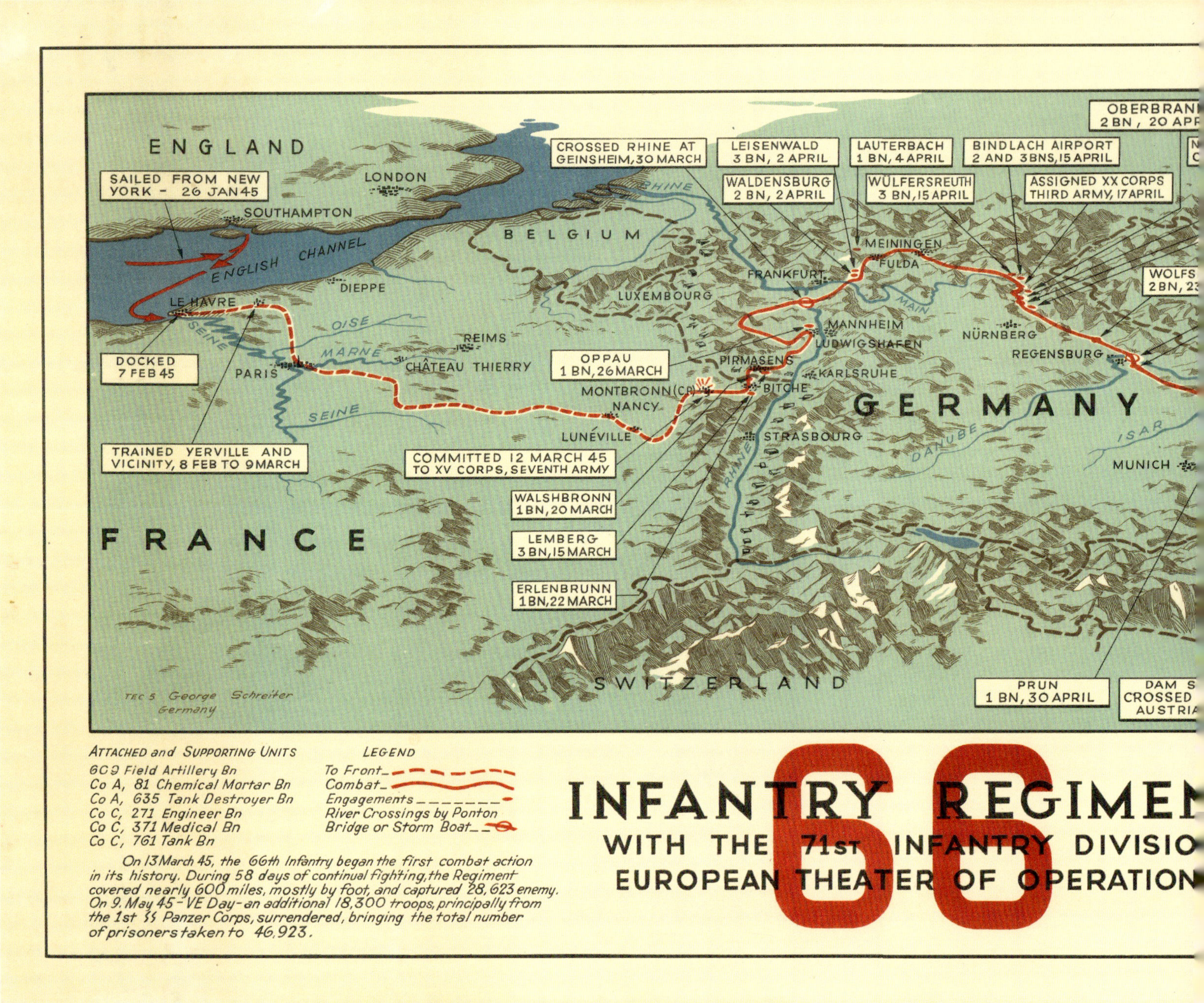

Schreiter (George): 66th Infantry Regiment with the 71st Infantry Division – European Theater of Operations

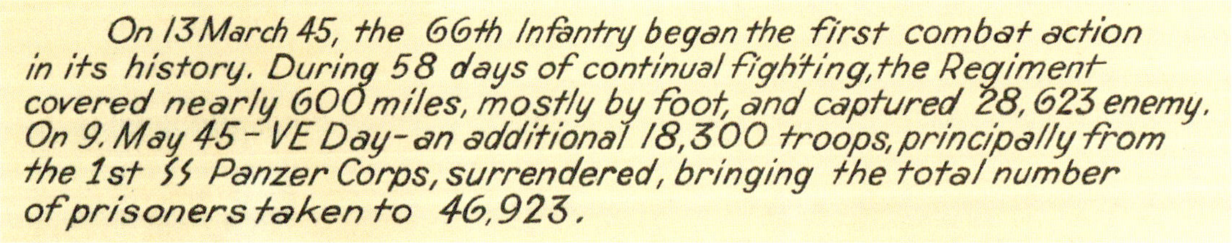

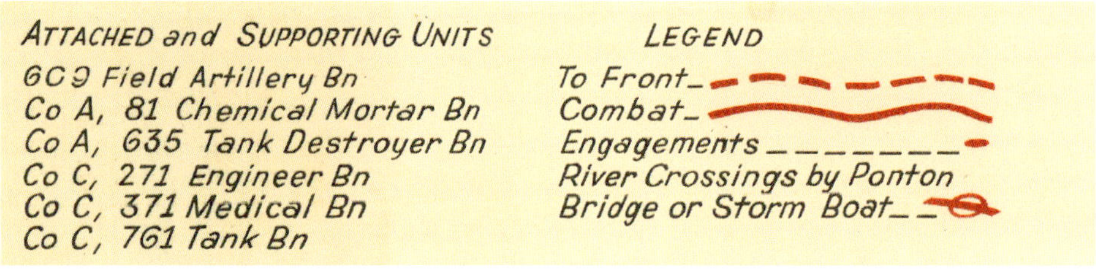

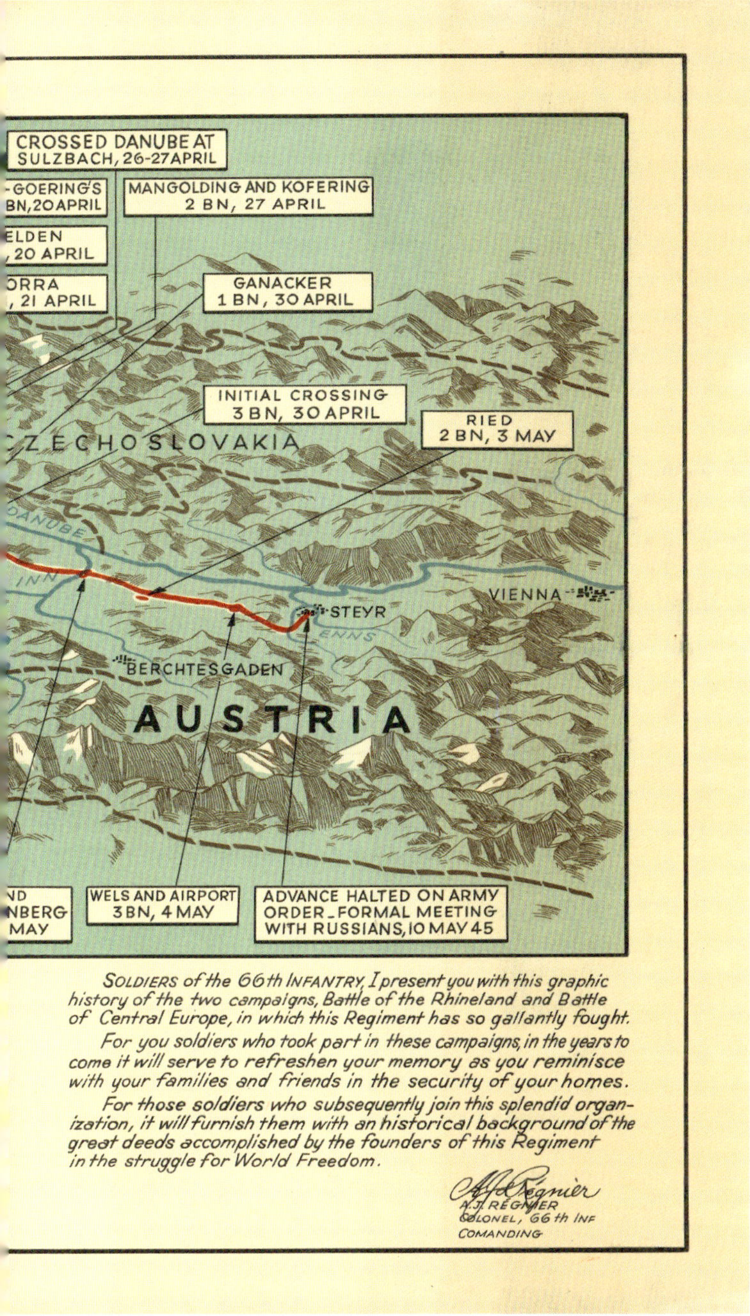

Schreiter (George): 66th Infantry Regiment with the 71st Infantry Division – European Theater of Operations

lithographed map, 1945, 30 x 16 in. (76 x 41 cm)

"Soldiers of the 66th Infantry, I present you with this graphic history of the two campaigns, Battle of the Rhineland and Battle of Central Europe, in which this Regiment has so gallantly fought."

The 71st Infantry Division, of which the 66th Regiment was part, had an unusual genesis. It was originally conceived in 1943 as a light division of 9,000 men supported by companies of African American mule drivers and hundreds of mules. Trained in mountain and jungle warfare, it was destined for the Pacific theatre. After testing manoeuvres indicated that light divisions were too small to be generally effective, it was retrained as a standard infantry division and prepared for redeployment on the newly opened European front.

Freshly trained and landing in France in February 1945, without having suffered to attrition of the D-Day Landings and the subsequent break-out, the 71st were ideally positioned to help drive the disordered German forces back into Germany itself. The 66th Regiment saw combat for the first time on 13 March and crossed the Rhine into Germany on 30 March.[247] The 71st Division advanced further east than any other U.S. Army formation and only came to a halt on 10 May 1945 when they met the Russians advancing from the opposite direction.

Among the supporting units attached to the 66th were the 81st Chemical Mortar Battalion and the 761st Tank Battalion. The chemical mortars were so-called because of their ability to fire chemical as well as incendiary and high explosive rounds.[248] The 761st Tank Battalion was a segregated battalion made up of African American soldiers.[249] In a belated recognition of their distinguished wartime record the battalion was awarded a Presidential Unit Citation by President Jimmy Carter in 1978.

George Schreiter was a Technician Fifth Grade (TEC 5), a technician rank addressed as corporal and as such nicknamed 'Tech Corporals'. These were eventually removed from the U.S. Army in 1948 but replaced by the Specialist rank in 1955. The dedication in the lower left of the map is by the 66th Infantry Regiment commander, Colonel A. J. Regnier.

[247] www.history.army.mil/html/forcestruc/cbtchron/cc/071id.htm; Pletcher, Capt. J. D., Nichols, Pfc. Norman., Coffman, Major Cameron and Tax, Cpl. Jerry, 1945, *The Seventy-First came to Gunskirchen Lager*, Gunskirchen.

[248] Despite its ban in the 1925 Geneva Protocol, both America and Britain maintained large supplies of chemical weapons, including chlorine and mustard gas, for use in retaliation against any potential German deployment of poison gas. The largest American stockpile of mustard gas, over 6,000 tons held at Deseret Chemical Depot in Utah, was only destroyed in 2012.

[249] General Patton addressed the men of 761st Battalion shortly before they went into combat for the first time, *"Men, you're the first Negro tankers to ever fight in the American Army. I would never have asked for you if you weren't good. I have nothing but the best in my Army. I don't care what colour you are as long as you go up there and kill those Kraut sons of bitches. Everyone has their eyes on you and is expecting great things from you. Most of all your race is looking forward to your success. Don't let them down and damn you, don't let me down!"* Wilson, Joe W., 1999, *The 761st "Black Panther" Tank Battalion in World War II*, Jefferson, NC, p. 53.

M. E. B.: Hell on Wheels 1942. 1943, 1944 – Second United States Armored Division

M. E. B.: Hell on Wheels 1942. 1943, 1944 – Second United States Armored Division

lithographed map, G.2 Section - 2nd A. D. - Passed by Censor, early 1945, 28 x 22 in. (71 x 56 cm)

This elegant American map celebrates the Second World War's achievements of the illustrious Second Armored Division, celebrated as 'Hell on Wheels'.

The upper part of the map shows the progress of the division from its landing in Normandy after D-Day to its entry into Germany. The lower panel illustrates its earlier involvement in the North African Campaign and the invasion of Sicily in 1942 and 1943. Dividing the two parts are the names of the eight generals who commanded the 2nd Armored in action, from Charles L. Scott, George S. Patton through to E.N. Harmon, all surrounded by a border of tank tracks.

The 2nd Armored was activated in July 1940, and had its first training at Fort Benning, Georgia. After taking part in training manoeuvres in the southern U.S. and special training off the Carolina coast under the Atlantic Fleet Amphibious Force, the first units were sent to North Africa in September 1942, followed by more training.[250] The division landed in French Morocco on 8 November 1942 and distinguished itself that same month under the command of Patton in the drive towards Casablanca during Operation Torch. The 2nd Armored later landed at Gela and Licata on the Sicilian coast in July 1943 as part of Operation Husky, and saw heavy combat around Palermo. In need of some R&R, and in preparation for Operation Overlord, the division arrived in England in late November 1943.

The most celebrated of 2nd Armored commanders was, of course, George Patton. One of the most dashing and brilliant Allied field commanders of the Second World War, Patton was notorious for his inspirational if obscenity-laden 'blood and guts' speeches to his men, his ivory-handled revolvers[251] and the infamous slapping of two shell-shocked soldiers in a hospital in Sicily.

[250] http://2ndarmoreddivision.org/history/

[251] Patton is often described as carrying 'pearl-handled revolvers' but as he stated to a reporter "Son, only a pimp in a cheap Louisiana whorehouse would carry a pearl-handled gun. These are ivory." See Perry, Milton; Parke, Barbara, 1957, *Patton and his Pistols*, Harrisburg PA

Patton was born in 1885, and after graduating from the U.S Military Academy at West Point, represented the United States in the Pentathlon at the Stockholm Olympics in 1912. In 1914 he joined the staff of General John J. 'Blackjack' Pershing during the expedition against Pancho Villa in Mexico.[252] In 1917 Captain Patton was the first man to be assigned to the newly established United States Tank Corps and was present at the Battle of Cambrai in France, the largest tank battle to date of the First World War. At the Meuse-Argonne Offensive in 1918, Patton was badly wounded in the leg while leading the American Tank Corps in one of the final battles of the war.

When Congress was finally persuaded to create an armoured force in 1940, Patton, a passionate advocate for tanks as the future of warfare, was transferred to Fort Benning and appointed Commanding General of the 2nd Armored Division. He led them and the entire Seventh Army into battle during Operation Torch and the invasion of Sicily. After the public relations disaster when he slapped a hospitalised soldier suffering from battle fatigue, he was relegated to an administrative role.[253] In 1944 immediately after D-Day, Eisenhower, recognising his supreme talents as a battlefield commander, gave him command of the Third Army, once more including his old outfit, the 2nd Armored.

Three days after D-Day, once several bridgeheads had been secured, the 2nd Armored landed on the Cotentin Peninsula. It participated in the fight for Carentan, broke through the St. Lô-Periers Line and fought off the notorious and much feared 2nd SS Panzer Division *Das Reich*, as it tried to relieve German forces trapped in the Normandy pocket.

Two and a half months later, in pursuit of the German Seventh Army, the division crossed the Seine west of Paris, and raced across France to become the first Allied troops to enter Belgium on 2 September 1944. Their attack on the Siegfried Line near Ubach led them to positions on the River Roer in November, from where they were redirected to take part in the Battle of the Bulge on the Meuse River. During Christmas 1944 they fought off the deepest German penetration in this winter counter-offensive in the Ardennes by the 5th Panzer Army, again coming up against the 2nd Panzer Division and this time virtually annihilating it.

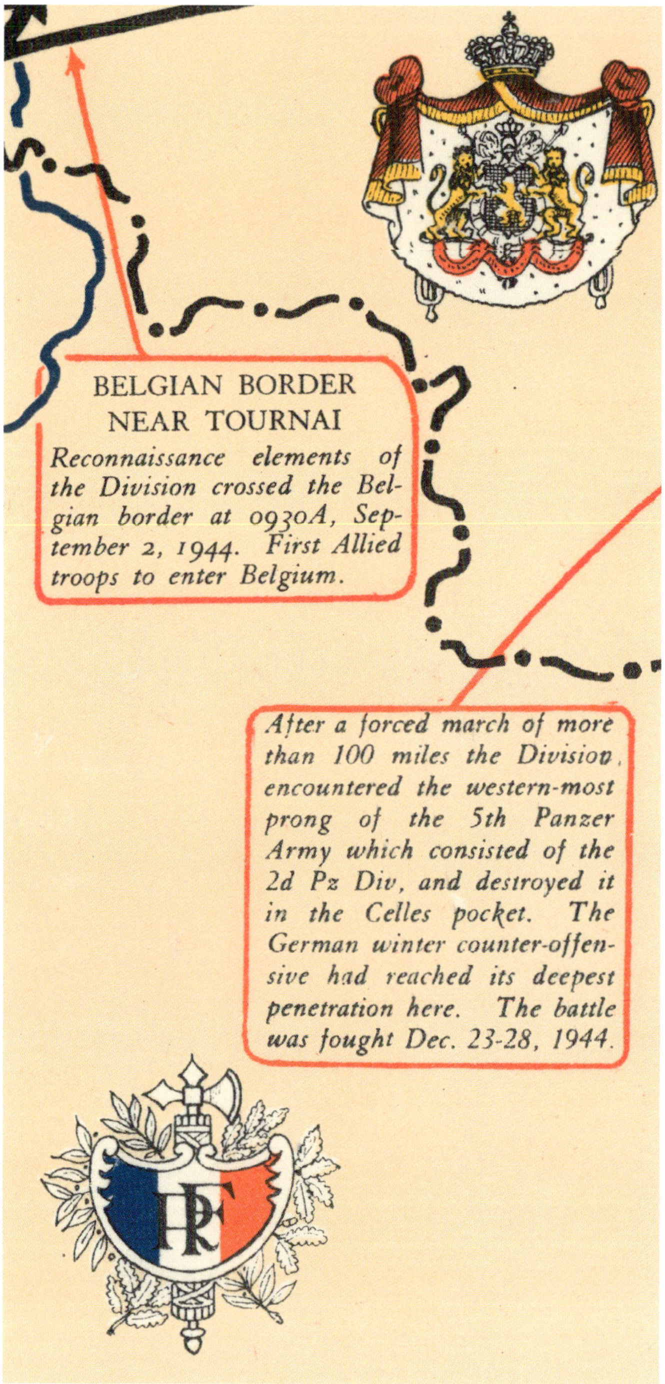

[252] In 1916, Patton, leading a force of tensoldiers and two guides in three Dodge touring cars, surprised a group of Pancho Villa's foragers, in what is considered the first motorised attack in the history of U.S. warfare.

[253] Atkinson, Rick, 2008, *The Day of Battle: The War in Sicily and Italy, 1943-1944*, pp. 147-148.

The constraints on publishing the routes of an active unit is reflected in the note "Passed by Censor". The American Office of Censorship was a powerful and, by necessity, secretive organisation controlling all information going into and out of the United States. Anything that could be of use to the enemy including the movements of the President, factory production figures, even weather reports, was ruthlessly supressed.[254]

"HELL ON WHEELS"

1942 • 1943 • 1944

This map was drawn and published early in 1945; shortly after the end of the Battle of the Bulge but before the 2nd Armored crossed the Rhine on 27 March. In July the Division became the first American unit to enter Berlin.[255]

On 9 December 1945, General Patton himself died twelve days after a minor traffic accident involving his staff car and a U.S. lorry in Germany. The unexplained nature of Patton's fatal wounds, particularly the baffling fracture he received to the back of the neck, have led to a proliferation of conspiracy theories, principally that he was killed by factions within the Soviet Union or even U.S. secret service critical of his fervent anti-communist stance.[256]

The 2nd Armored was one of very few heavy armoured divisions to maintain its organisation throughout the Second World War. It saw 50 more years of active service, stationed in Germany on several occasions during the Cold War and in 1970 was commanded by George Patton IV, the first time in U.S. history a division had ever been commanded both by father and son. The Division saw its last action in Operation Desert Storm during the First Gulf War in 1990-91. It was decommissioned as part of the restructuring after the collapse of the Soviet Union, and finally, after a series of re-designations, the renowned Hell on Wheels ended its service in 1995.

[254] Throughout much of the war, radio sports commentators were even banned from describing the weather conditions under which any sporting contests were being conducted. Sweeney, Michael S., 2001, *Secrets of Victory: The Office of Censorship and the American Press and Radio in World War II*. Chapel Hill, N.C.

[255] http://2ndarmoreddivision.org/history/

[256] Philip Curtis heard the story behind Patton's fatal crash from the late Michael Moore, son of Major General Cecil Ray Moore, Eisenhower's Chief Engineer for the European Theatre of Operations. The relatively recent surge in the number of conspiracy theories relating to Patton's death arose because the traffic accident occurred between slow-moving vehicles and Patton was the only person seriously hurt. He received a non-life threatening blow to the forehead, consistent with being thrown forward onto the glass partition but what ultimately killed him was the seemingly inexplicable breaking of the vertebra at the base of his neck, similar to that delivered by a sap or karate chop according to some of the more sensationalist theories.

It was shortly after Patton's death that General Moore was assigned a driver from the pool allocated to the American top brass, from whom Patton's own driver had also been drawn on the day of the accident. His driver confided to him that it wasn't so much the crash itself that killed Patton; rather it was the crate of whisky on the car's rear shelf which shot forward and caught him on the back of the neck. The embarrassing if not actually incriminating case of whisky was promptly removed from both the car and the subsequent reports, leaving a mystery that has delighted conspiracy theorists ever since.

Drury (John G.): Okinawa Mem-O-Map

Drury (John G.): Okinawa Mem-O-Map

lithographed map, Mem-O-Map Co., 1945, 10 x 13 in. (26 x 33 cm)

This essentially light-hearted map was designed as a commemorative souvenir for the soldiers and marines stationed on Okinawa.

Okinawa was seen by both the Americans and Japanese as the last major stepping stone for an assault on mainland Japan. The Japanese were determined to defend it quite literally to the death. Over 100,000 troops were dug in amongst hundreds of heavily fortified bunkers and caves all over the island. The battle to take Okinawa proved to be the most brutal of the whole Pacific war.

To capture the island, the Americans dispatched a vast armada of naval ships and landing craft in the largest amphibious landing of the Pacific Theatre. Despite relentless kamikaze attacks, progress for the American soldiers and marines was at first surprisingly easy with rapid territorial gains. This changed dramatically when they came across the prepared defences, and there followed some of the grimmest fighting of the war. Using high explosives, flamethrowers and savage hand-to-hand combat, they were forced to clear the island cave by cave, bunker by bunker with the Japanese fighting with fanatical desperation, contesting every inch. The Americans used bulldozers to seal thousands of Japanese soldiers into their labyrinth of underground defences which became their tombs. When all hope was lost, many of the defenders committed suicide rather than face the shame of capture.

After 82 days of combat the death toll was horrendous. The Japanese lost over 150,000, many of them civilians. The Americans suffered 65,000 casualties including Lieutenant General Simon Bolivar Buckner, commander of the 10th Army and the highest ranking American officer to be killed by enemy fire.[257] The extensive divisional cemeteries are all marked on the map.

Another significant casualty was the revered war correspondent, Ernie Pyle, shot through the head on the island of Ie Shima and whose memorial is also shown.

The huge number of casualties, combined with the dogged Japanese defence in the face of certain defeat, was a significant factor in the decision to drop the atomic bombs on Hiroshima and Nagasaki just six weeks later.

This Mem-O-Map was designed to appeal to the large numbers of U.S. military personnel stationed on Okinawa after the battle in the build-up for the anticipated invasion of Japan. The no doubt limited opportunities for recreation for the garrison forces on the devastated island are also represented, including a vignette of a horse-riding G.I. captioned 'Small Horses'. 'Uma-harase' was a traditional form of Okinawan racing on a miniature breed of native horse. This breed is now extinct, as all the stallions were castrated during the First World War in order to encourage the breeding of larger warhorses and it seems unlikely that any survived into the 1940s.[258]

[257] Beevor, Antony, 2012, *The Second World War*, London, pp. 707-708.

[258] The Okinawa Zoo is attempting to reintroduce traditional uma-harase racing using imported miniature Korean horses. See: *Ryukyu Shimpo*, 4 February 2013.

This form of map drawn by John Drury, who served with 214th Ordnance Battalion, must have proved popular as the Mem-O-Map Company went onto publish further examples for the Philippines, Hawaii and Europe.

Drury (John G.) Mem-O-Map of Europe, 1945

Drury (John G.) Mem-O-Map of the Philippine Islands, 1945

Bullock (Leslie George): United Nations Map of the World

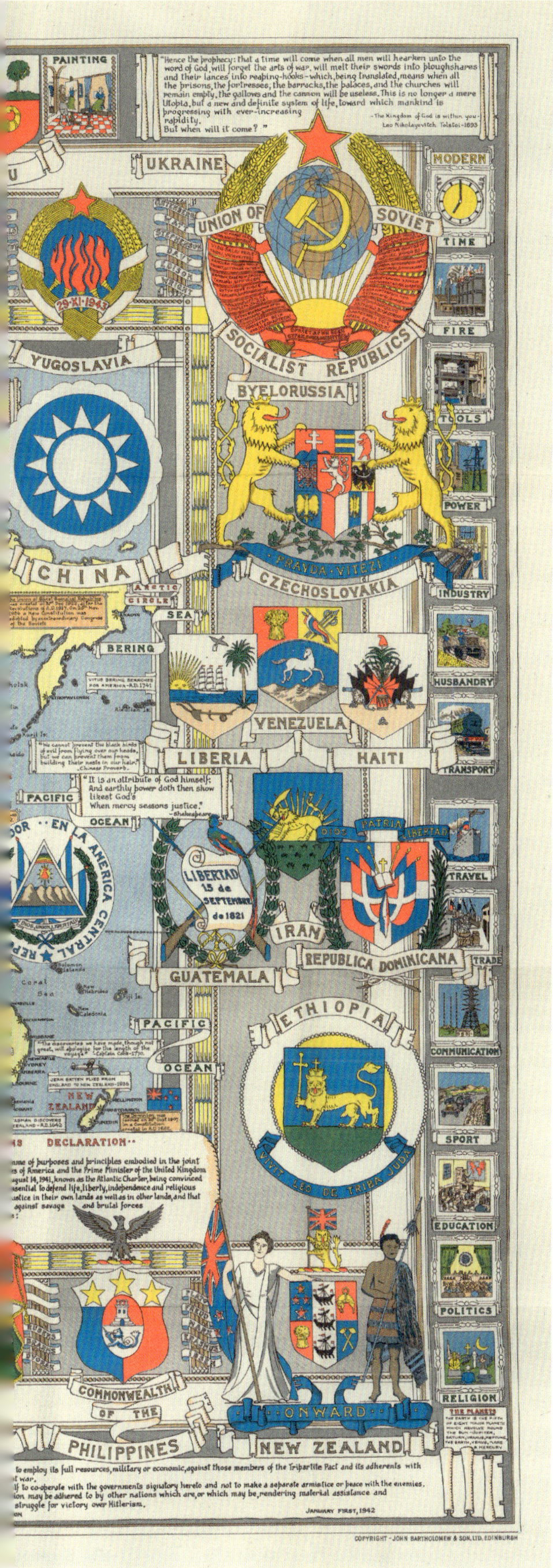

Bullock (Leslie George): United Nations Map of the World

lithographed map, John Bartholomew & Son, Ltd., Edinburgh, 1948,
36 x 26 in. (91 x 66 cm)

This linen-backed folding map is a lavish visual celebration of the establishment of the United Nations.

Following on from Macdonald Gill's map celebrating the ideals expressed in the Atlantic Charter, John Bartholomew & Sons in Edinburgh published this map, created by L. G. Bullock, to commemorate the creation of the United Nations in 1945. Gill's map, while looking forward to a better future and even suggesting how to achieve it, was drawn while the Second World War still raged. Bullock's post-war map reflects a more optimistic time and is decorated with symbols of hope and security.[259]

The coats of arms of the founder nations decorate the border with prominence given to the United States, the Soviet Union and Britain. Emphasizing that the UN was a creation of the victors of the Second World War, the former Axis powers are notable for their absence. Italy joined in 1955, Japan the following year and the two divided Germanys only joined in 1973.

The map is a teeming cornucopia of decorative details. In the lower corners of the map itself are the texts of the Atlantic Charter and the United Nations Declaration. Also quoted are Euripides, the New and Old Testaments, Shakespeare, Captain Cook, Abraham Lincoln, Tolstoy, and many more explorers, writers and philosophers. The European discoveries, along with the first circumnavigation, flights, etc. are indicated by text and vignettes.

While the purpose of this map, the promotion of peace, reflects that of the UN itself, it cannot avoid all references to the previous world conflict. A small mushroom cloud is shown over Japan while Germany itself remains unnamed and unmarked, reflecting the current uncertainty as to what post-war form it would take.

The designer of the map, L. G. Bullock, also produced pictorial maps of England & Wales, Scotland, Ireland and the Children's Map of London. They were all published by John Bartholomew & Son, the most important Scottish publisher of maps during the last half of the nineteenth-century and the first part of the twentieth-century. The son, also called John, was a founding member of the Scottish Geographical Society and named the continent of Antarctica. John George Bartholomew, the grandson of the founder of the company, became Geographer and Cartographer to King George V.[260]

[259] Whitfield, Peter, 2010, *A Life in Maps*, London, p. 132.

[260] Worms, Laurence and Baynton-Williams, Ashley, 2011, *British Map Engravers*, London, pp. 52-53.

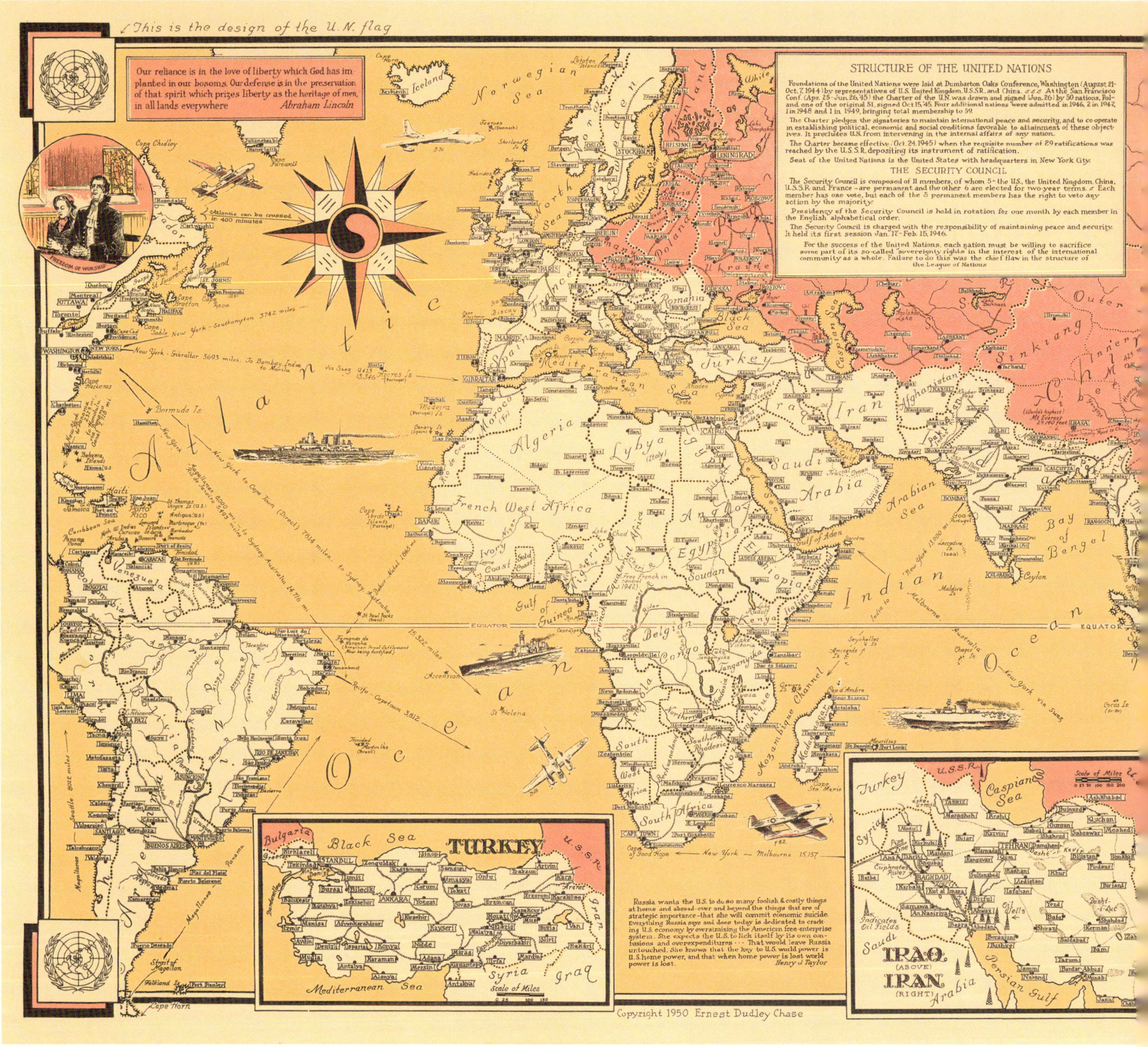

Dudley Chase (Ernest): A Factual and Pictorial Map of World Freedom

Dudley Chase (Ernest): A Factual and Pictorial Map of World Freedom

lithographed map, Ernest Dudley Chase, Winchester, 1950, 35 x 15 in. (89 x 38 cm)

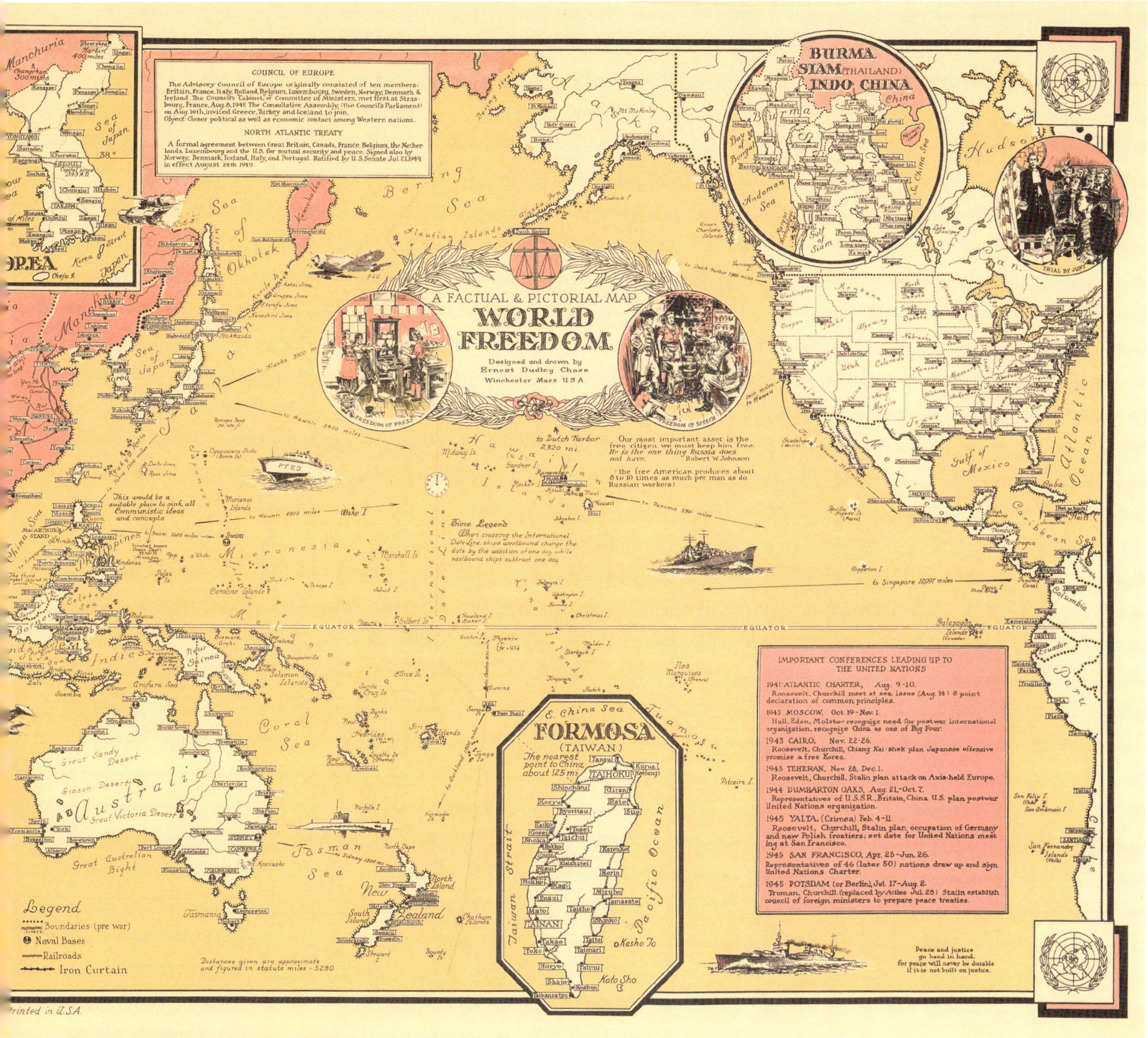

It would be nice, if disingenuous, to finish with a map that celebrates worldwide peace. Sadly the history of the second half of the twentieth-century and current events teaches us that conflict, whether ethnic, religious, political or philosophical, is ever with us.

Dudley Chase's *World Freedom*, so similar to his earlier *Total War Battle Map* published in 1942, vividly shows how rapidly the Allies of the Second World War had been transformed into the implacable enemies of the Cold War. The map colouring now starkly contrasts the lands controlled by Communist Russia and China and the rest of the world.

The vignettes of ships and airplanes are similar to the earlier map but the images of weapons have been replaced by insets of disputed territories, explanatory text panels and idealistic scenes portraying the freedoms of speech, of the press, of worship and trial by jury. The areas of potential conflict presciently include Korea, Indochina and Vietnam, and Iraq and Iran.

Dudley Chase makes his own feelings clear with his choice of a quote by Henry J. Taylor included in the lower part of the map: "*Russia wants the U.S. to do so many foolish and costly things at home and abroad – over and beyond the things that are of strategic importance – that she will commit economic suicide. Everything Russia says and does today is dedicated to cracking U.S. economy by overstraining the American free-enterprise system...*" Elsewhere Chase states that the free American produces about eight to ten times as much per man as do Russian workers.[261] Additionally the note placed beside "the greatest known Ocean depth 35,400 ft, Mindanao Deep" off the east coast of the Philippines which Dudley Chase had earlier suggested "This would be a good place to sink the Jap fleet" has been replaced by "This would be a suitable place to sink all Communistic ideas and concepts." [262]

Both Bullock's *United Nations Map* and Gill's *Atlantic Charter Map* anticipate that the defeat of the Axis Powers will lead to an era of peace with UN negotiated diplomacy replacing combat. By 1950 this sense of optimism, together with the high-minded principles behind the creation of the U.N., have already degenerated into a new form of conflict between superpowers.

[261] This is a gross overestimation as Russian productivity, man for man, during this period is reckoned at between 25 per cent and 40 per cent of the American. See: Kouwenhoven, Remco, 1997, *A Comparison of Soviet and US Industrial Performance: 1928-90*, *Research Memorandum GD-29*, Groningen Growth and Development Centre.

[262] In 1951, the year after this map was published, HMS *Challenger* surveyed the Mariana Trench and revealed it to be even deeper at over 36,000 ft.

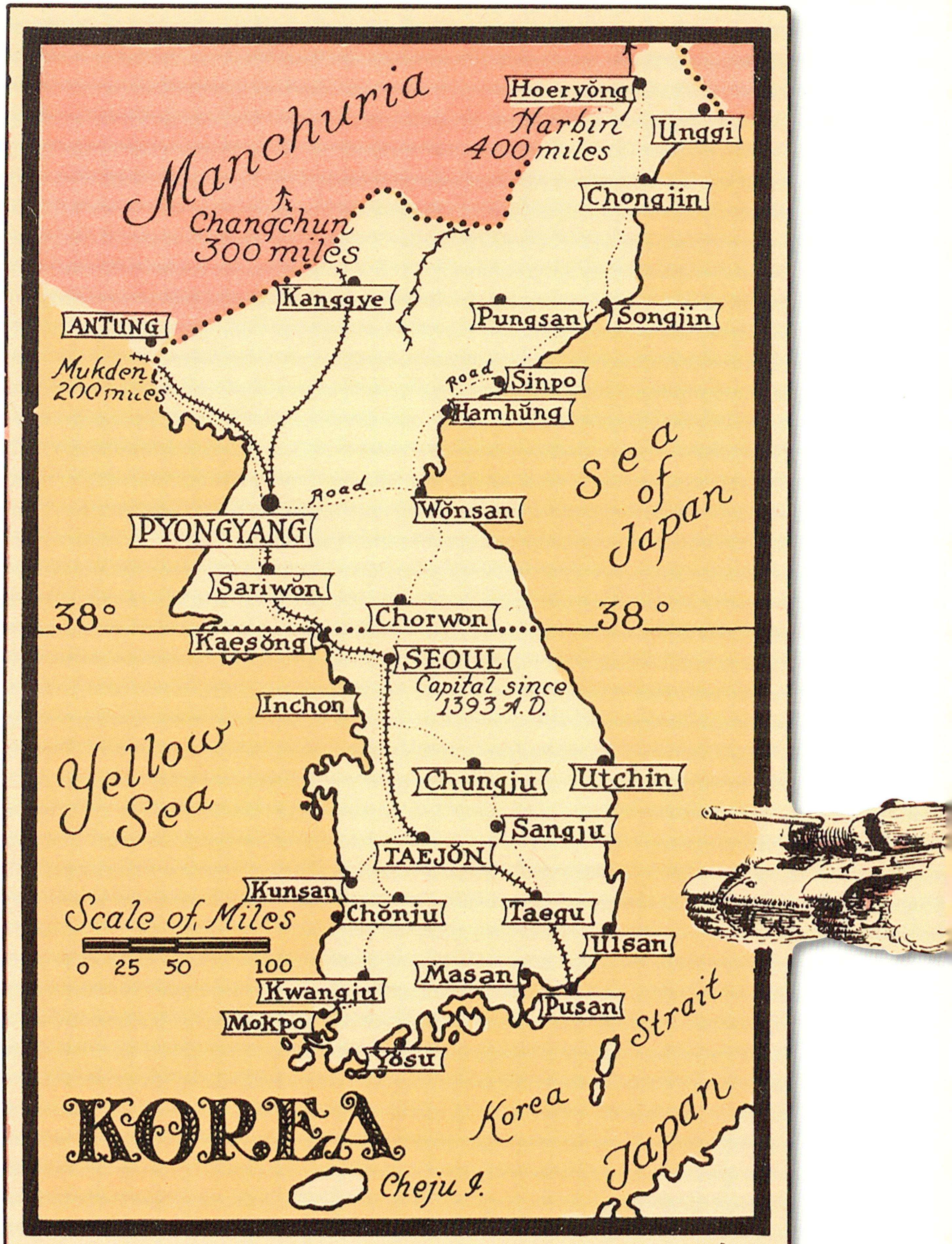
Manchuria
Changchun 300 miles
Hoeryŏng
Harbin 400 miles
Unggi
Chongjin
Kanggye
ANTUNG
Pungsan
Songjin
Mukden 200 miles
Road
Sinpo
Hamhŭng
Sea of Japan
Road
PYONGYANG
Wŏnsan
Sariwŏn
38°
Chorwon
38°
Kaesŏng
SEOUL
Capital since 1393 A.D.
Inchon
Yellow Sea
Chungju
Utchin
Sangju
TAEJŎN
Kunsan
Scale of Miles
Chŏnju
Taegu
Ulsan
0 25 50 100
Masan
Kwangju
Pusan
Mokpo
Korea Strait
Yŏsu
KOREA
Japan
Cheju I.

MAROCCO
SIFTON, PRAED & Co. LTD.
The Map House
67, St. James's Street, LONDON. S.W.
Sierra Leone and Liberia.
ORDNANCE SURVEY
ENGLAND & WALES
1 INCH TO A MILE
SHEET
PRICE THREE SHILLINGS.
C. SMITH & SON'S
TAPE
INDICATOR
MAP OF
LONDON
SIFTON, PRAED & Co. LTD.
67, ST. JAMES'S STREET, S.W.
1903
Environs of London
(Geological)
SHEET 9
BARTHOLOMEW'S ROAD MAP
OF
ENGLAND AND WALES
Scale: 4 Miles to an inch
SCOTLAND
INDEX TO SHEETS
SIFTON PRAED AND CO. LTD.
67, ST. JAMES'S STREET, LONDON, S.W.
SHEET 125
REIGATE &
TUNBRIDGE WELLS
Sicily
East
SOUTH BRAZIL
SIBERIAN RAILWAY
Ruanda
Urundi.
SHEET 116
Western Front.
Ascot
Windsor
53 Cadogan Square
S.W.
"Our Country is the World"
SIFTON PRAED AND CO. LTD.
67, St. James's Street, London, S.W.
Mapsellers, Geographical and Military
Booksellers and Stationers

THE MAP HOUSE

Sifton Praed & Co Ltd

In 1907, Alfred Sifton and Francis Praed founded Sifton Praed & Co Ltd as a book and map publisher and seller. The company trading as The Map House was based at 67 St James's Street in London.

The Map House soon became a familiar landmark in its own right and was one of the fixtures that trainee taxi drivers had to know if they were to pass "The Knowledge" and become an official London cabby. Daphne du Maurier even included The Map House as the setting for a crucial plot twist in her collection *The Birds & Other Stories*, *"I had gone into the Map House, in St James's to buy myself some half-dozen books to read on that long thrash across the Atlantic – a journey one took with certain qualms in those days, the Titanic tragedy still fresh in memory – and there were Victor and Anna, poring over maps, which they had spread out over every available space."* The Map House is still as welcoming to enthusiastic browsers today as it was then.

While selling maps of all parts of the world, The Map House was particularly noted for its production of large scale folding maps. As well as covering the most distant parts of the world, it was also renowned for catering to a new type of map buyer: the first car owners. Map sets were produced in leather-bound cases for the use of the discerning motorists setting off to explore the British Isles and beyond. Each case came with its own miniature wheeled map measurer and compass for calculating relative distances on the folding paper maps.

Photograph by Jonathan Potter

Of course The Map House didn't only provide maps for car owners and in 1938 in an article on a travel book, *Third Class, How to Travel without Being Rich*, the *Tablet* refers to *"The Map House, St James Street, S.W. – the fountainhead of all pleasures on foot."*

The Map House supplied maps to travellers, explorers, colonial administrators and all branches of the armed services. Ernest Shackleton bought his maps of Antarctica from us, and Winston Churchill, his maps of the Far East. In 1920 we received a Royal Warrant from the Prince of Wales, for supplying the future Edward VIII with maps of the Western Front during the Great War.

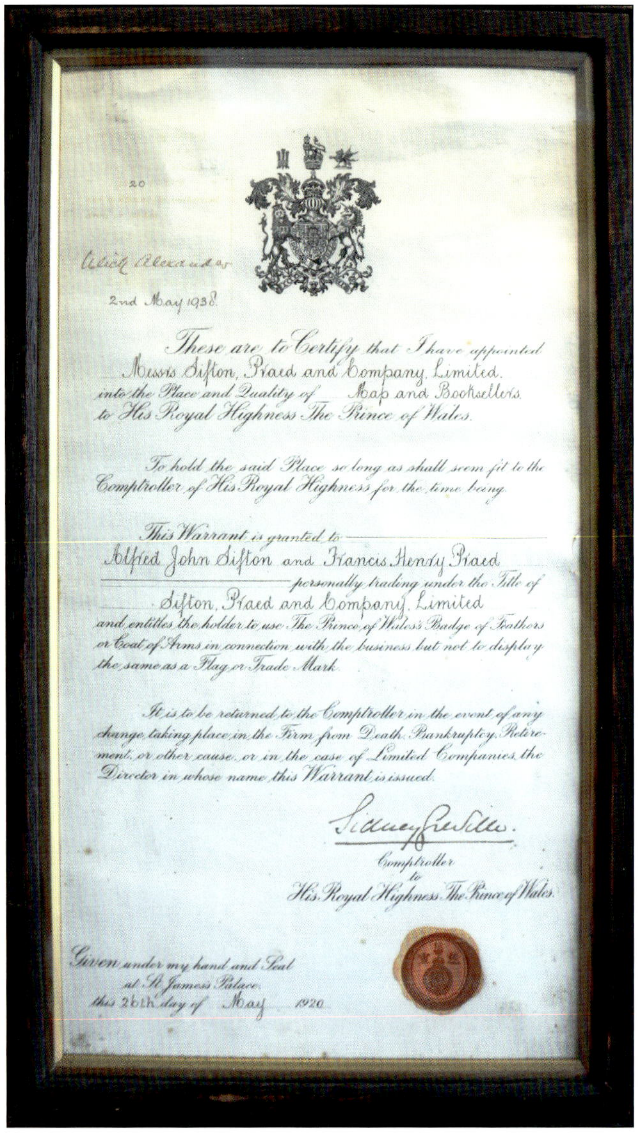

20

2nd May 1938

These are to Certify that I have appointed Messrs Sifton, Praed and Company Limited into the Place and Quality of Map and Booksellers to His Royal Highness The Prince of Wales.

To hold the said Place so long as shall seem fit to the Comptroller of His Royal Highness for the time being.

This Warrant is granted to Alfred John Sifton and Francis Henry Praed personally trading under the Title of Sifton, Praed and Company Limited and entitles the holder to use The Prince of Wales's Badge of Feathers or Coat of Arms in connection with the business but not to display the same as a Flag or Trade Mark.

It is to be returned to the Comptroller in the event of any change taking place in the Firm from Death, Bankruptcy, Retirement, or other cause, or in the case of Limited Companies the Director in whose name this Warrant is issued.

Comptroller to His Royal Highness The Prince of Wales

Given under my hand and Seal at St James's Palace this 26th day of May 1920

The strong emphasis on selling military maps was also reflected in a long list of titles published by Sifton Praed on military mapping, instruction books on how best to carry out military surveys, as well books on historical campaigns.

As official licensed distributor of the maps for the War Office during both the First and Second World Wars, The Map House was not only instrumental in the distribution of maps detailing the ongoing conflict but also played a part in publishing and distribution of propaganda maps such as those illustrated in this book.

In more peaceful times productions have ranged from sporting maps for the fox hunter and fly-fisherman to guides for the part-time farmer or the visitors to the Lake District or the Caribbean.

In the early 1970s Sifton Praed ceased to publish or print any more maps and books and continued its transformation into one of the first dedicated antiquarian map shops. The Map House has been supplying superb examples of cartographic art to collectors and institutions including the British Library and the Library of Congress ever since.

After nearly seventy years in St James's, the Map House moved to its present position at 54 Beauchamp Place in Knightsbridge in 1973. It is now home to the finest and most comprehensive selection of original antique and vintage maps, globes and atlases offered for sale anywhere in the world. Maps of all periods are represented, ranging from rare and often unique cartographic masterpieces from the dawn of printing in the fifteenth century to fascinating records of Victorian exploration and 20th-century pictorial flamboyance.

More recently, the Sifton Praed legacy of publishing books on maps has been revived, with the first edition of *The Mapping of Antarctica* being produced in 2012 and now with *War Map* in 2016. The Map House is proud and delighted to be returning this aspect of the business over a century after it published its first works.

www.themaphouse.com
maps@themaphouse.com
+44 (0)207 589 4325

SIFTON, PRAED & CO. PUBLISHING

The Map House

54, Beauchamp Place, LONDON, SW3 1NY

Select Bibliography

Anderson, Ross, 2004, *The Forgotten Front, The East Africa Campaign, 1914–1918*, London.

Army Ordnance, vol. 15-16, American Defense Preparedness Association, 1934.

Asahi Shimbun Shashi, *Asahi Shimbun Sha, 1990–1995*, Tokyo and Osaka.

Atkinson, Rick, 2008, *The Day of Battle: The War in Sicily and Italy, 1943-1944*, New York.

Barber, Peter & Harper, Tom, 2010, *Magnificent Maps*, London.

Baridon, Laurent, 2011, *Un Atlas Imaginaire, Cartes Allégoriques et Satiriques*, Paris.

Bar-Zohar, Michael, 2001, *Beyond Hitler's Grasp: The Heroic Rescue of Bulgaria's Jews*, Avon, MA.

Beevor, Antony, 2012, *The Second World War*, London.

Biddle, Tami Davis, 2004, *Rhetoric and Reality in Air Warfare*, Princeton, NJ.

Boyce, Robert W. D. (ed.), 1998, *French Foreign and Defence Policy, 1918-1940: The Decline and Fall of a Great Power*, London & New York.

Broadbent, Harvey, 2005, *Gallipoli: The Fatal Shore*, Camberwell.

Bryant, Mark, 1994, *Dictionary of British Cartoonists and Caricaturists, 1730-1980*, Aldershot.

Bryant, Mark, 2000, *Dictionary of Twentieth Century British Cartoonists and Caricaturists*, Aldershot.

Bryant, Mark, 2006, *World War I in Cartoons*, London.

Bryant, Mark, 2005, *World War II in Cartoons*, London.

Bryant, Mark, 2011, *The World's Greatest War Cartoonists and Caricaturists, 1792-1945*, London.

Bryars, Tim and Harper, Tom, 2014, *A History of the 20th Century in 100 Maps*, London.

Buchbender, Ortwin, 1978, *Das tönende Erz: Deutsche Propaganda gegen die Rote Armee im Zweiten Weltkrieg*, Stuttgart.

Budiansky, Stephen, 2004, *Air Power: The Men, Machines, and Ideas that Revolutionized War, from Kitty Hawk to Gulf War II*, London.

Burdon, Elisabeth, 2009, *Macdonald Gill: The "Wonderground Map" of 1913 and its influence, IMCoS Journal, vol. 116*, London.

Burtness, Robert A. 2012, *The Santa Barbera B-24 Disasters: A Chain of Tragedies Across Air, Land & Sea*, Stroud, Gloucestershire.

Cahm, Eric, 2014, *The Dreyfus Affair in French Society and Politics*, London.

Cannon, John, 1997, *The Absent-Minded Beggar, Gilbert and Sullivan News*, March 1997, vol. 11, no. 8, London.

Carringer, Robert. L., 1979, *The Jazz Singer*, Madison, WI.

Chappell, Mike, 1996, *Army Commandos 1940–1945*, London.

Chelminski, Rudolph, 1997, *The Maginot line, Smithsonian Magazine, June 1997*, Washington, DC.

Churchill, Winston, 1951, *Closing the Ring, History of the Second World War, vol.5*, London.

Churchill, Winston, 1981, Speech at the Guildhall, 9 November 1914, *The Oxford Library of Words and Phrases*, Oxford.

Clark, George B., 2013, *The American Expeditionary Force in World War I: A Statistical History, 1917-1919*, Jefferson, NC.

Collectif de Recherches de l'Université et Musées Lausannois, 1986, *19-39: la Suisse Romande entre les Deux Guerres*, Lausanne.

Commander-in-Chief Pacific Fleet to Commander-in-Chief Fleet, 28.6.42, NHHC, *Battle of Midway*: 4-7 June 1942 Action Reports, *F-2042*.

Costello, John & Hughes, Terry, 1977, *The Battle of the Atlantic*, London.

Creelman, James, 1901, *On the Great Highway, the Wanderings and Adventures of a Special Correspondent*, Boston (re-issued online by Cardinal Books: www.cardinalbook.com/creelman/highway/iso8859/)

Darracott, Joseph, 1989, *A Cartoon War, World War Two in Cartoons*, London.

Darracott, Joseph & Loftus, Belinda, 1972, *Second World War Posters*, London.

Deighton, Len, 1979, *Blitzkrieg*, London.

Dobbins, James, Poole, Michele A., Long, Austin & Runkle, Benjamin, 2008, *After the War: Nation Building from FDR to George W. Bush*, Santa Monica, CA.

Dorléac, Laurence Bertrand, 2008, *Art of the Defeat: France 1940-1944*, Los Angeles, CA.

Doyle, William, 2002, *The Oxford History of the French Revolution*, Oxford.

Eicher, John H. & Eicher, David J., 2001, *Civil War High Commands*, Palo Alto, CA.

Emery, Elizabeth, 2011, Albert Robida, Medieval Publicist, in Utz, Richard and Emery, Elizabeth, *Cahier Calin: Makers of the Middle Ages. Essays in Honor of William Calin*, Kalamazoo, MI.

Engdahl, William, 1992, *A Century of War: Anglo-American Oil Politics and the New World Order*, San Diego, CA.

Eversdijk, Nicole, 2010, Kultur als politische Werbemittel, *Studien zur Geschichte und Kultur Nordwesteuropas, Band 21*, Münster.

Fabyanic, Lt. Col. Thomas A., 1985, *The French Air Force of 1940: Was It Defeated by the Luftwaffe or by Politics,* Air University Review September-October, Pennsylvania.

Fairchild, Byron, 1960, Decision to land United States Forces in Iceland, 1941, *Command Decisions*, United States Army Center of Military History, Washington, DC.

Ferrell, Robert H. 2012, *America's Deadliest Battle: Meuse-Argonne, 1918*, Lawrence, KS.

Flying Magazine, June 1934, Chicago, IL.

Gailey, Harry A. 1996, *The War in the Pacific: from Pearl Harbor to Tokyo Bay*, Novato, CA.

Gardner, Hall, 2015, *The Failure to Prevent World War I: The Unexpected Armageddon*, Farnham.

Grehan, John & Mace, Martin (eds.), 2014, *Western Front 1917-1918: Despatches from the Front*, Barnsley.

Hannegan, Barry, December 1997, Ezra C. Stiles: A Re-introduction, *PHLF News*, no. 148, Philadelphia.

Harmon, Katharine, 2003, *You are Here: Personal Geographies and Other Maps of the Imagination*, Princeton.

Harrison, Richard Edes, 1943, *The War of the Maps, Saturday Review of Literature*, New York.

Harrison, Richard Edes, 1944, *Look at the World: The Fortune World Atlas for World Strategy*, New York.

Heckman, Wolf, 1981, *Rommel's War in Africa*, New York.

Henry, Mark & Walsh, Stephen, 2003, *The U.S. Army of World War I (Men-at-Arms)*, Oxford.

Herzstein, Robert Edwin, 2005, *Henry R. Luce, Time, and the American Crusade in Asia*, Cambridge.

Hilliard Atteridge, A. 1929, *A History of the 17th (Northern) Division*, Glasgow.

Hooton, E. R., 2007, *Luftwaffe at War: Gathering Storm 1933-39, vol. 1*, London.

Hosch, William L., 2009, *World War I: People, Politics, and Power*, New York.

Hoxson, Walter L. (ed.), 2002, American Diplomacy in the Second World War, *The American Experience in World War II, vol. 8*, New York and London.

Hughes, Edan Milton, 1989, *Artists in California 1786-1940, II*, Sacramento, CA.

Humphries, M. O. & Maker, J. 2013, Der Weltkrieg: 1914 *The Battle of the Frontiers and Pursuit to the Marne, Germany's Western Front: Translations from the German Official History of the Great War, I, part 1*, Waterloo, Canada.

Jukes, Geoffrey, 2002, *The First World War, The Eastern Front 1914-1918*, Oxford.

Kahn, David, 1967, *The Codebreakers*, New York.

Kaplan, Philip, 2014, *World War Two at Sea: The Last Battleships*, Barnsley.

Kaufmann, J. E. & Kaufmann, H. W., 2006, *Fortress France: The Maginot Line and the French Defenses in World War II*, Mechanicsburg, PA.

Kaufmann, J. E. & Kaufmann, H. W., 2010, *The American GI in Europe in World War II*, Mechanicsburg, PA.

Keegan, John, 1999, *The First World War*, London.

Killen, John, 2003, *The Luftwaffe: A History*, Barnsley.

Klee, Ernest, 2007, *Das Kulturlexikon zum Dritten Reich. Wer war was vor und nach 1945*, Frankfurt am Main.

Knox, MacGregor, 1999 [1982], *Mussolini Unleashed, 1939–1941: Politics and Strategy in Fascist Italy's Last War*, Cambridge.

Kodansha, 1983, *Kodansha Encyclopedia of Japan*, Tokyo and New York.

Kouwenhoven, Remco, 1997, A Comparison of Soviet and US Industrial Performance: 1928-90, *Research Memorandum GD-29*, Groningen Growth and Development Centre.

Landers, James, 2010, *The Improbable First Century of Cosmopolitan Magazine*, Columbia, MO.

Leitz, Christian, 2001, *Nazi Germany and Neutral Europe during the Second World War*, Manchester.

Lenman, Robin, 1997, *Artists and Society in Germany, 1850-1914*, Manchester.

Life Magazine, 17 February 1941, New York.

Life Magazine, 15 June, 1942, New York.

Low, David, 1946, *Years of Wrath – A Cartoon History*, London.

Manning, Martin J., Wyatt, Clarence R. (eds.), 2010, *Encyclopaedia of Media and Propaganda in Wartime America*, Santa Barbara, CA.

Marchand, Susanne E., 2009, *German Orientalism in the Age of Empire: Religion, Race and Scholarship*, Cambridge.

Marder, Arthur J. 1976, *Operation 'Menace': the Dakar expedition and the Dudley North Affair*, Oxford.

Matanle, Ivor, 1995, *World War II: The 50thAnniversary of the Allied Victory*, Sydney.

Merrill, John C. and Fisher, Harold A. 1980, *The World's Great Dailies: Profiles of Fifty Newspapers*, New York.

Mollo, Andrew, 1981, *The Armed Forces of World War II*, New York.

Nasaw, David, 2013, *The Chief: William Randolph Hearst,* Boston.

Nead, Lynda, 2007, *The Haunted Gallery*, New Haven and London.

Nedd, Andrew M., Segodniashnii Lubok: Art, War, and National Identity in James, Pearl (ed.), 2010, *Picture This: World War I Posters and Visual Culture*, Lincoln, NE.

Nevins, Allan & Weitenkampf, Frank, 1944, *A Century of Political Cartoons, Caricature in the United States from 1800 to 1900*, New York.

Oehrig, Cristen, *Civil Affairs in World War II*, Center for Strategic and International Studies, Washington, DC.

Olson, James Stuart & Shadle, Robert (eds.), 1991, *Historical Dictionary of European Imperialism*, Westport, CT.

Osterwalder, Marcus, 1983, *Dictionnaire des Illustrateurs (1800-1914) - Illustrateurs, Caricaturistes et Affichistes*, Paris.

Pati, Budheswar, 1991, *India and the First World War*, New Delhi.

Personalities in the Realm of Art, Macdonald Gill, F.R.I.B.A, *Town and Country News*, 29 September 1933.

Pletcher, Capt. J. D., Nichols, Pfc. Norman., Coffman, Maj. Cameron and Tax, Cpl. Jerry, 1945, *The Seventy-First Came to Gunskirchen Lager*, Gunskirchen, Austria.

Ponsonby-Fane, Richard, 1962, *Sovereign and Subject*, Kyoto.

Pravda, 29.3.35, Moscow.

Proctor, Ben, 2007, *William Randolph Hearst: The Later Years, 1911-1951*, Oxford.

Proctor, John, 2012, *American Resolve and the Art of War: A Study and Application of Military Tactics*, Bloomington, IN.

Riqueti, Comte de Mirabeau, Honoré Gabriel, 1788, *De la Monarchie Prussienne sous Frédéric le Grand*, London.

Röel, John, 2014, *Kaiser Wilhelm II: A Concise Life*, Cambridge.

Roth, Eugen, 1957, *100 Jahre Humor in der Deutschen Kunst*, Hannover.

Ryukyu Shimpo, 4 February, 2013, Okinawa.

Sanders, Paul, 2005, *The British Channel Islands under German Occupation, 1940-1945*, Jersey.

Schmidt, Klaus Werner, 1982, Überwachungs-Reglementierungen im Ersten Weltkrieg (1914-1918), in Fischer, Heinz-Dietrich (ed.), *Deutsche Kommunikationskontrolle des 15. bis 20. Jahrhunderts*, Munich.

Schneer, James A., 2015, *Mikasa - An Illustrated History of the Last Pre-dreadnought Battleship*, Morrisville, NC.

Schneider, James, 2011, *Guerrilla Leader: T. E. Lawrence and the Arab Revolt*, New York.

Seldes, George, 1929, *You Can't Print That*, New York.

Semenoff, Vladimir, 1907, *The Battle of Tsushima*, London

Seth, Robert, 1960, *Two Fleets Surprised. The Story of the Battle of Cape Matapan, Mediterranean, March, 1941*, London.

Sheffield, Gary and Todman, Gary (eds.) 2007, *Command and Control on the Western Front: The British Army's Experience 1914–18*, London.

Short, Neil, 2004, *Germany's West Wall – The Siegfried Line*, Oxford.

Skinner, H. T. & Stacke, H. Fitz M. 1922, Principal Events 1914-1918, *History of the Great War Based on Official Documents*, London

Smele, Jonathan D. 1996, *Civil War in Siberia: The Anti-Bolshevik Government of Admiral Kolchak, 1918-1920*, Cambridge.

Smith, Colin, 2009, *England's Last War Against France: Fighting Vichy, 1940-1942*, London.

Soucacos, Panayotis, 2015, *Satirical Maps*, Athens.

Stasi, Jean-Charles, 2015, *L'Incroyable Raid de Granville*, Bayeux.

Stedman, Michael, 2001, *Advance to Victory 1918: Somme*, Barnsley.

Strachan, H. 2001, To Arms, *The First World War I*, Oxford.

Sweeney, Michael S. 2001, *Secrets of Victory: The Office of Censorship and the American Press and Radio in World War II*. Chapel Hill, NC.

Syrett, David, 1994, *The Defeat of the German U-boats: The Battle of the Atlantic*, Columbia, SC.

'Tantamount to a declaration of war': The National Archives, Kew, *ADM 199/391.*

Taylor, A. J. P. & Mayer, S. L. (eds.), 1974, *A History of World War Two*. London.

The Argus, Melbourne, 14 January 1941.

The Daily Mail, December 1899, London.

The National Archives, 2005, *Allied Propaganda in World War II: The Complete Record of the Political Executive*, London.

'The orders of His Majesty's Government': The National Archives, Kew, *ADM 399/192.*

Thrift, Tim, 1945, *A Meticulous Maker of Maps*, Boston.

Time Magazine, 20 July 1936, New York.

Tomasevich, Jozo, 2002, *War and Revolution in Yugoslavia, 1941-1945: Occupation and Collaboration*, Stanford, CA.

Tooley, R. V. 1963, Map Collector's Series, No. 1: Geographical Oddities, *Map Collector's Circle*, London.

Tucker, Spencer (ed.), 2005, *The Encyclopaedia of World War I: A Political, Social, and Military History*, Santa Barbara, CA.

Tyng, Sewell, 1935, *The Campaign of the Marne 1914*, New York.

United States Government Printing Office, 1947, *Building the Navy's Bases in World War II: History of the Bureau of Yards and Docks and the Civil Engineer Corps, 1940-1946, vol. 2*, Washington, DC.

Vogel, Walter, 1941, Die Organisation der Amtlichen Presse- und Propagandapolitik des Deutschen Reiches von den Anfängen unter Bismarck bis zum Beginn des Jahres 1933, *Sonderheft der Monatschrift für internationale Zeitungsforschung, Zeitungswissenschaft (16), 8/9*, Berlin.

Wells, William Bittle, 1904, *Pacific Monthly, v. XI*, Portland, OR.
Westwell, Ian & Cove, Dennis (eds.), 2002, *History of World War I, vol. 3*, New York.
Whitfield, Peter, 2010 (A), *A Life in Maps*, London.
Whitfield, Peter, 2010 (B), *The Image of the World*, London.
Wilson, Jeremy, 1990, *Lawrence of Arabia*, New York.
Wilson, Joe W., 1999, *The 761st "Black Panther" Tank Battalion in World War II*, Jefferson, NC.
Wood, Robert J., 1937, Early Days of Benny Havens, *The Pointer*, February 26.
Wood, Walter, 1911, *North Sea Fishers and Fighters*, London.
Worms, Laurence and Baynton-Williams, Ashley, 2011, *British Map Engravers*, London.
Youngblood, Norman, 2006, *The Development of Mine Warfare: A Most Murderous and Barbarous Conduct*, Greenwood.
Zaalberg, Thijs Brocades, 2006, *Soldiers and Civil Power: Supporting or Substituting Civil Authorities in Modern Peace Operations*, Amsterdam.
Zeman, Zbynek, 1978, *Selling the War, Art and Propaganda in World War II*, London
Ziemke, Earl F. 2004, *The Red Army, 1918-1941: From Vanguard of World Revolution to America's Ally*, London.

INTERNET RESOURCES:

2nd Armored Division Association (http://2ndarmoreddivision.org/history/)
Artindex.nl
(www.artindex.nl/noordbrabant/default.asp?id=6&num=047190008701901014309007700183096050163 1&relt=&limit=&print=yes)
Comic Shop Lambiek (www.lambiek.net/artists/v/von_baumgarten_eugen.htm)
Der Bildindex der Kunst und Architektur - Image Index for Art and Architecture (www.bildindex.de/obj14050115.html#|0)
Doughboy Center (www.worldwar1.com/dbc/stmihiel.htm)
Engle Fine Art Auctions (www.auktionshaus-engel.de/texte/kuearch/Eddelbuttel,%20Richard.htm)
Exeter University: Bombing, States and Peoples in Western Europe 1940-1945, *The Bombing of Italy 1940-45* (http://centres.exeter.ac.uk/wss/bombing/italy.htm)
Global Security.org (www.globalsecurity.org/military/world/europe/skoda-pilsen.htm)
www.fdonaldblake.com/
www.hammondmap.com/
www.uboat.net/allies/technical/asdic.htm
Imperial War Museum Catalogue (www.vads.ac.uk/large.php?uid=51971)
Kleukens Archive (http://kleukens-archiv.de/flash/LP/teil1/index.html)
Library of Congress (www.loc.gov/loc/lcib/0305/hammond.html)
National Archives of Australia (www.naa.gov.au/collection/fact-sheets/fs195.aspx)
Publishers Weekly (www.publishersweekly.com/pw/by-topic/industry-news/publisher-news/article/63089-global-publishing-leaders-2014-kodansha.html)
Rickard, J. 2007, *First Battle of Picardy, 22-26 September 1914*: (www.historyofwar.org/articles/battles_picardyI.html)
Scarborough Evening News, 24 October 1904: (www.scarboroughsmaritimeheritage.org.uk/adoggerbank.php)
State Library of Massachusetts (http://mastatelibrary.blogspot.co.uk/2009/05/ernest-dudley-chase-worldview-in-maps.html)
The D-Day Museum (www.ddaymuseum.co.uk/d-day/)
The Historical Museum of Fort Missoula Catalogue (http://fortmissoulamuseum.org/WWII/detail.php?id=403)
The Naval History and Heritage Command, at the Washington Navy Yard: www.history.navy.mil/faqs/faq36-4.htm
The official West Point website: www.usma.edu/wphistory/SitePages/Home.aspx
University of Pittsburgh (http://beta.worldcat.org/archivegrid/data/181596538, http://digital.library.pitt.edu/cgi-bin/f/findaid/findaid-idx?type=simple;c=ascead;view=text;subview=outline;didno=US-PPiU-ais200008)
U.S. Army Center of Military History (www.history.army.mil/html/forcestruc/cbtchron/cc/071id.htm)
War casualties in Nijmegen 1939-1945 (www.oorlogsdodennijmegen.nl/persoon/manche/acf6f25c-2ccb-4633-bed6-1ff958e0ec0a)
West-Point.Org – The West Point Connection (www.west-point.org/greimanj/west_point/wpl.htm)
Yale University, October 18, 1907, *Laws of War: Opening of Hostilities (Hague III)*, Avalon Project at Yale Law School (http://avalon.law.yale.edu/20th_century/hague03.asp)

MAROCCO
SIFTON, PRAED & Co. LTD.
The Map House
67, St James's Street, LONDON. S.W.
Sierra Leone and Liberia.
ORDNANCE SURVEY
ENGLAND & WALES
1 INCH TO A MILE
SHEET
PRICE THREE SHILLINGS.
C. SMITH & SON'S
TAPE INDICATOR
MAP OF
LONDON
SIFTON, PRAED & Co. LTD.
67, ST JAMES'S STREET, S.W.
Environs of London
[Geological]
SHEET 9
BARTHOLOMEW'S ROAD MAP
OF
ENGLAND AND WALES
Scale: 4 Miles to an inch
SCOTLAND
INDEX TO SHEETS
SIFTON PRAED AND CO. LTD.
67, ST. JAMES'S STREET, LONDON, S.W.
SHEET 125
REIGATE &
TUNBRIDGE WELLS
Sicily
East
SOUTH BRAZIL
SIBERIAN RAILWAY
Ruanda
Urundi.
SHEET 116
"Our Country is the World."
SIFTON PRAED AND CO. LTD.
67, St. James's Street, London, S.W.
Mapsellers, Geographical and Military
Booksellers and Stationers